THE GLITCHES OF REALITY

PART TWO: SIMULATION THEORY

Elias Verdan

Heinrich Wilson Publishing

Contents

Introduction: The Programmed Universe

In 1973, science fiction author Philip K. Dick stood before an audience in France and made a startling claim: "We are living in a computer-programmed reality, and the only clue we have to it is when some variable is changed, and some alteration in our reality occurs." At the time, his words seemed like the paranoid delusions of a drug-addled writer. Today, they sound prophetic.

Fifty years later, simulation theory—the hypothesis that our universe might be an artificial construct running on some advanced computational substrate—has moved from fringe speculation to serious philosophical discourse. Nobel Prize-winning physicists debate it. Tech billionaires fund research into it. University philosophy departments teach courses on it. The question is no longer whether simulation theory is worth discussing, but whether we can afford to ignore its implications.

What changed between Dick's speech and today? The answer is simple: technology caught up to imagination. We became simulators ourselves.

We created video games with procedurally generated worlds spanning millions of square miles. We built virtual reality systems that can trick the human brain into believing digital spaces are physical. We developed neural networks that learn, adapt, and solve problems in ways that mirror biological intelligence. We watched computational power double every eighteen months for decades, approaching levels that once seemed impossibly distant. And somewhere in that progression, a question became unavoidable: if we can create increasingly sophisticated simulations, and if technology keeps advancing exponentially, where does this trajectory end?

The uncomfortable answer: it ends with simulations indistinguishable from base reality. And if such simulations are possible—not just theoretically, but practically achievable by sufficiently advanced civilizations—then the mathematics suggests something startling. We're almost certainly living in one.

This is not a comfortable thought. It violates our intuitive sense that the world around us is solid, fundamental, real in some absolute sense. It challenges the basic assumption underlying all human endeavor: that what we experience is the ground floor of existence rather than some intermediate level in a vast hierarchy of nested realities. It raises questions about free will, consciousness, meaning, and purpose that most of us would prefer not to confront.

But comfort has never been a reliable guide to truth.

When Plato described prisoners in a cave, watching shadows on a wall and mistaking them for reality, he was working with metaphor. When ancient Hindu philosophers spoke of Maya—the veil of illusion obscuring true reality—they were describing spiritual concepts divorced from physical mechanisms. When the Wachowskis created The Matrix in 1999, they were making entertainment, not scientific predictions. But today, when we discuss the possibility that our reality is simulated, we're not speaking metaphorically or spiritually. We're making a concrete, testable hypothesis about the fundamental nature of existence.

The difference is crucial. Simulation theory isn't mysticism dressed in technological language. It's a serious attempt to explain why the universe behaves the way it does, why physical laws take the precise mathematical forms they do, why quantum mechanics operates according to rules that seem designed for computational efficiency, and why consciousness has such a strange relationship with physical reality.

In Part One of The Glitches of Reality, we explored quantum immortality—the hypothesis that consciousness cannot experience its own termination. We examined how the many-worlds interpretation of quantum mechanics suggests a multiverse of constantly branching possibilities, and how observer selection effects might create the subjective experience of impossible survival. We looked at cases where people should have died but didn't, where near-misses seem statistically improbable, where reality appears to preserve conscious observers against overwhelming odds.

That exploration raised as many questions as it answered. If consciousness has this strange relationship with quantum measurement, what does that tell us about the nature of consciousness itself? If reality branches based on observation, what is doing the observing, and where does that observation occur? If subjective experience is preserved across quantum splits, where is that preservation mechanism located? And most fundamentally: what kind of universe operates according to rules that seem specifically designed to create and maintain conscious observers?

Part Two addresses these questions by examining what might be the most comprehensive explanation for reality's strangeness: that we exist inside a simulation, a programmed construct running on computational substrate we cannot directly perceive or access. This hypothesis, properly understood, doesn't dismiss or contradict quantum immortality. Instead, it provides a framework that makes quantum immortality's implications less mysterious and more comprehensible.

A simulated reality designed to generate and maintain conscious experience would naturally implement mechanisms ensuring those experiences continue. The observer-dependent nature of quantum mechanics makes more sense if reality is rendered based on observation rather than existing independently of observers. The fine-tuning of physical constants becomes less miraculous if they're simply parameters set during initialization rather than random cosmic accidents. The mathematical elegance of physical laws becomes less surprising if those laws are literally code—algorithms governing how the simulation operates.

Simulation theory is not a rejection of science. It's a hypothesis that takes scientific findings seriously and follows them to their logical conclusions. It doesn't require faith in supernatural forces or mystical energies. It doesn't demand you abandon empirical investigation or rational inquiry. Instead, it asks you to consider the possibility that the universe operates on computational principles, that consciousness is a phenomenon emerging from information processing, and that the reality you experience might be one level removed from whatever constitutes base existence.

This book will explore simulation theory systematically and rigorously. We'll begin with the foundational philosophical arguments—particularly Nick Bostrom's statistical reasoning for why we're probably simulated rather than existing at the base level of reality.

We'll examine the computational nature of the universe itself, looking at how physical laws take mathematical forms and why reality seems to operate according to principles that mirror information processing.

We'll investigate quantum mechanics through the lens of computational rendering, examining how phenomena like wave-function collapse, entanglement, and superposition might reflect processing limitations or optimization strategies. We'll confront the fine-tuning problem—the fact that fundamental constants are calibrated to impossible precision—and consider whether this suggests intentional design rather than cosmic luck.

We'll also examine the evidence that troubles skeptics: the glitches, anomalies, and strange experiences that some interpret as artifacts of imperfect programming. Déjà vu. The Mandela Effect. Synchronicities that defy statistical probability. We'll approach these phenomena seriously without abandoning critical thinking, asking what they might tell us if simulation theory is correct, and what alternative explanations might account for them if it's not.

Then we'll confront the counterarguments. Simulation theory has serious philosophical problems that honest inquiry cannot ignore. Critics raise valid objections about infinite regress, falsifiability, computational requirements, and the assumption that consciousness can emerge from computation. We'll examine these objections thoroughly, acknowledging that simulation theory might be wrong and exploring what that wrongness would mean.

Finally, we'll move beyond abstract philosophy to practical implications. If we're simulated, who might be running the simulation and why? Does it matter for ethics, meaning, and purpose? Can we have genuine free will inside a deterministic program? What changes in how we should live if we accept that reality might be programmed? And perhaps most importantly: how do we maintain intellectual humility and emotional equilibrium while contemplating questions this fundamental?

This is not a book that will tell you definitively whether we're simulated. That answer, if it exists, remains beyond our current reach—perhaps permanently, by design. Instead, this is a book about taking a difficult idea seriously, following it wherever it leads, and seeing what insights emerge when we question our most basic assumptions about reality.

The simulation hypothesis is either true or false. There is no middle ground. But the question itself—the mere fact that we can ask it coherently, that it's become a legitimate topic of scientific and philosophical investigation—tells us something profound about where human civilization has arrived. We've reached a point in our development where we must seriously consider that we might be characters in someone else's story, processes in someone else's program, thoughts in someone else's mind.

What we do with that possibility matters.

Let's begin.

Chapter One

The Simulation Hypothesis—Core Arguments

The simulation hypothesis rests on a deceptively simple argument: if it's possible to create realistic simulations of conscious beings, and if civilizations tend to create many such simulations, then statistically speaking, you're almost certainly living in one right now.

This argument, formalized by philosopher Nick Bostrom in 2003, doesn't require exotic physics or spiritual assumptions. It's pure probability. And once you understand it, it's difficult to dismiss.

Let's start with the basics. Imagine a future civilization—maybe our own descendants, maybe aliens, maybe something we can't currently imagine—that has access to tremendous computational power. Not just supercomputers or quantum processors, but technology so advanced it makes our current capabilities look like stone tools. This isn't science fiction speculation. It's extrapolation. Computing power has doubled roughly every eighteen months for decades. If that trend continues for even a few centuries, the computational resources available to future civilizations become almost incomprehensible.

What would such a civilization do with that power? Among many things, they might run simulations. Not simple video games or weather models, but detailed, high-fidelity simulations of entire worlds, complete with conscious beings who believe they're real. They might run these simulations for research—to study history, test social theories, understand how consciousness emerges. They might run them for entertainment, the way we play The Sims or Civilization but with actual conscious characters. They might run them for reasons we can't imagine, the way an ant can't understand why humans build particle accelerators.

Here's where the math gets interesting. If one advanced civilization creates simulated realities, they probably don't create just one. They create millions, or billions, or trillions. Why? Because once you have the capability, the marginal cost of running additional simulations approaches zero. It's like downloading a file—copying it a million times takes barely more resources than copying it once.

Now count the realities. You have one base reality—the "real" universe where the simulators live. And you have potentially billions or trillions of simulated realities they've created. If you're a conscious being asking "Am I real or simulated?" the odds overwhelmingly favor simulated. It's not even close. For every person in base reality, there might be a billion simulated people who feel equally real, think equally clearly, and have no way of knowing they're simulated.

This is Bostrom's core insight, and it's genuinely unsettling. The argument doesn't depend on whether we're currently being simulated. It depends only on whether such simulations are possible in principle. If they are, the sheer number of simulated realities drowns out base reality like a single voice in a stadium crowd.

But wait. Maybe advanced civilizations don't create these simulations. Maybe they can't, or won't, or find it unethical. Bostrom addresses this too, presenting what he calls the "trilemma"—three possibilities, at least one of which must be true.

The first possibility: civilizations like ours go extinct before reaching the technological capability to run realistic simulations. Maybe nuclear war gets us. Maybe climate change. Maybe artificial intelligence destroys its creators. Maybe there's some universal filter—a challenge so difficult that essentially no civilization survives to become simulator-gods. If this is true, we're not simulated because there's no one around to simulate us. But it also means our future is probably bleak. We're likely headed for extinction before we reach technological maturity.

The second possibility: advanced civilizations could create simulations but choose not to. Maybe they find it morally reprehensible to create conscious beings who don't know they're simulated. Maybe they lose interest in such things, the way humans stopped burning witches not because it became impossible but because our values changed. Maybe there's some advanced-civilization consensus that ancestor simulations—replicas of earlier evolutionary stages—are taboo. If this is true, we're not simulated because our descendants or parallel civilizations decided such simulations are wrong. But this possibility seems less likely the more you think about it. It requires essentially all advanced civilizations to agree, forever, never to run simulations. One rogue actor, one curious scientist, one entertainment company—that's all it takes to create billions of simulated realities. Universal agreement is hard to maintain.

The third possibility: we are living in a simulation. Right now. This conversation, these words, your thoughts while reading them—all of it rendered by some computational substrate beyond our perception. If simulations are possible and civilizations create them, this becomes the statistically likely answer. Not certain, but probable. Very probable.

Bostrom doesn't claim to know which possibility is true. He's simply presenting the logic: at least one of these three must be correct. Either we go extinct, or we choose not to simulate, or we're simulated. Pick your uncomfortable truth.

Most people's intuitive reaction is to reject the third possibility. "I feel real," they say. "The world is obviously real. I'm not a computer program." But this reaction misunderstands what simulation means. A simulated person would feel exactly as real as you do. Their thoughts would be genuine thoughts. Their experiences would be genuine experiences. The simulation wouldn't be fake in any meaningful sense—it would simply be implemented on different substrate than you assume.

Think about dreams. While dreaming, you believe the dream world is real. You interact with dream characters, experience dream emotions, make dream decisions. Only upon waking do you realize it was generated by your sleeping brain. But during the dream, it felt completely real. Your dream-self had no way of detecting that the entire experience was constructed internally.

Now imagine a dream so sophisticated that it never ends, that maintains perfect consistency, that includes billions of other dreamers all dreaming together in the same shared space. At what point does the distinction between "dream" and "reality" become meaningless? At what point is the simulation so complete that it simply is reality, just implemented on different substrate?

This brings us to what Bostrom calls "substrate independence"—the idea that consciousness doesn't require biological neurons. If the pattern of information processing matters rather than the physical material doing the processing, then consciousness can run on silicon, or quantum processors, or whatever technology the simulators use. Your thoughts would be exactly as real on artificial substrate as on biological neurons. The substrate doesn't matter. The pattern does.

This is a crucial point that many people miss. Simulation theory doesn't claim your experiences are fake or illusory. It claims they're real experiences implemented on computational substrate. The difference is profound. It's not saying you don't exist. It's saying you exist as information processing rather than as fundamental physical matter.

Consider a character in a video game advanced enough to be conscious—genuinely self-aware, capable of thought and feeling. Is that character's suffering less real because it runs on a computer? Is their joy less meaningful? If they ponder their existence, arrive at insights, form relationships, pursue goals—are these activities somehow diminished by being implemented in code rather than neurons?

The answer isn't obvious. And that ambiguity is where simulation theory gets philosophically interesting.

The substrate independence thesis challenges our intuitions about what counts as "real." We naturally assume that biological brains are special, that neurons firing in meat-space have some quality that distinguishes them from transistors switching in silicon. But this assumption might be nothing more than biological chauvinism—the prejudice that our particular implementation of consciousness is the only valid one.

Consider a thought experiment. Imagine a neuron in your brain. Now imagine replacing it with a tiny artificial device that performs exactly the same function—receives the same inputs, produces the same outputs, maintains the same connections to neighboring neurons. Would you notice? Would your consciousness flicker or change? Most neuroscientists would say no. One artificial neuron among billions wouldn't make a detectable difference.

Now replace another neuron. Then another. Continue this process, one neuron at a time, until your entire brain is artificial but functionally identical to the original. At what point did you stop being you? At what point did consciousness disappear? If the answer is "never"—if you remained conscious throughout because the functional pattern was preserved—then consciousness is substrate independent. It runs on the pattern, not the material.

This thought experiment, often called the "Ship of Theseus problem" when applied to consciousness, suggests that what matters isn't what you're made of but how that material is organized and what processes it implements. If true, this opens the door to consciousness existing in radically different substrates—including simulated ones.

But even if consciousness is substrate independent, does that make simulation likely? Here's where the technological trajectory argument comes in.

Look at the progress we've made in just seventy years of computing. In 1950, computers filled entire rooms and could barely perform basic calculations. Today, you carry exponentially more processing power in your pocket. Video games now feature worlds with realistic physics, dynamic weather, procedural generation creating billions of unique planets, and artificial intelligence controlling thousands of characters simultaneously. Virtual reality tricks your visual cortex into believing digital spaces are physical. Neural networks learn to write, draw, compose music, and engage in conversation.

We're nowhere near conscious simulations yet. Current AI, despite impressive capabilities, shows no signs of genuine self-awareness or subjective experience. But we've achieved this level in less than a century. Now project forward not another century, but a thousand years. Ten thousand. A million. If civilization survives and technology continues advancing, where do we end up?

The gap between our current capabilities and conscious simulations might be vast, but it's not obviously impossible. We see no physical laws preventing it, no fundamental barriers declaring "consciousness cannot be simulated." It might be difficult. It might require computational resources we can't currently imagine. It might take civilizations thousands of years to achieve. But difficult isn't the same as impossible.

And here's the crucial point: it only needs to be possible once. One civilization, somewhere in the vast universe, across all of cosmic time, needs to develop the capability and the inclination to run ancestor simulations. Just one. Because once they do, they create not one simulated reality but millions. And each of those simulated realities, if they develop their own technology, might create simulations of their own. The number of simulated realities quickly explodes to numbers that make the base reality statistically irrelevant.

Think of it like a pyramid scheme, but for realities. One base reality at the top. That reality creates a thousand simulations. Each of those creates a thousand more. Just three levels deep, you have a billion simulated realities for every base reality. The odds of being in base reality become microscopic.

This reasoning has convinced some serious thinkers that we're probably simulated. Elon Musk has said the odds we're in base reality are "one in billions." Many physicists and philosophers, while more cautious in their language, admit the argument is difficult to dismiss. It's not proof, but it's a strong probabilistic case built on reasonable assumptions.

But what counts as a simulation? This is where confusion often creeps in.

When most people hear "simulation," they think of The Matrix—humans in pods, physical bodies in a real world, consciousness jacked into a fake virtual reality. But that's not what simulation theory proposes. In the strong version of the hypothesis, there are no bodies in pods. There is no "real" physical world separate from the simulation. The simulation is all there is at your level of reality. You are not a physical being experiencing a simulated world. You are a pattern of information that has always existed only within the simulation.

This distinction matters. The Matrix scenario has humans escaping the simulation to find the "real" world. But in actual simulation theory, there's nowhere to escape to. The simulation isn't a false layer over reality—it is your reality, completely and entirely. Breaking out doesn't mean finding your true body. It means ceasing to exist, because you are a pattern that exists only within the simulation's computational substrate.

This is harder to grasp intuitively, but it's the version that makes philosophical sense. If you're simulated, you're not a real person being fed fake experiences. You're a genuine pattern of consciousness implemented in code rather than atoms. Your experiences aren't fake. They're real experiences happening to a real conscious entity. The substrate—the underlying "stuff" implementing you—is simply one level removed from what you perceive as physical reality.

Some people find this comforting. If the simulation is reality—if there's no "real" world being hidden from you—then you're not being deceived. You're simply discovering what your reality is made of, the same way discovering atoms doesn't make tables and chairs less real.

Others find it existentially disturbing. It means the fundamental layer you thought you understood—atoms, molecules, cells, neurons—might be rendered images of some deeper computational process. It means physical laws might be arbitrary choices made by programmers rather than fundamental truths. It means the universe's rules could be changed, the simulation could be paused, or it could be terminated entirely, wiping out everything you know without warning or recourse.

But here's a question worth sitting with: would it matter? If you learned tomorrow with absolute certainty that you're simulated, what would actually change? You'd still experience the same thoughts, feel the same emotions, face the same choices. The people you love would still matter. Your goals would still be worth pursuing. The sunset would still be beautiful, even if it's rendered rather than "real" in the way you once assumed.

Simulation theory doesn't make life meaningless. It just changes the metaphysics—the underlying nature of what existence is. And for day-to-day living, metaphysics might matter less than we think.

Still, the question haunts us. Are we real or rendered? Base reality or recursive simulation? Fundamental or derivative?

The math suggests an answer. But math isn't certainty. And even if we're probably simulated, "probably" leaves room for doubt—and for hope, depending on which possibility you find more troubling.

Chapter Two

The Computational Universe

The universe speaks mathematics. This isn't poetic language—it's an observation that has puzzled scientists and philosophers for centuries. Physical laws can be expressed as elegant equations. The motion of planets follows precise mathematical curves. Light behaves according to formulas that predict its behavior with stunning accuracy. Even the apparent chaos of quantum mechanics submits to mathematical description, probabilistic but perfectly consistent.

Why should this be true? Why should the universe operate according to mathematical principles at all?

The standard answer is that mathematics is the language we use to describe reality, a human invention that happens to map well onto physical phenomena. But this explanation feels insufficient when you examine how unreasonably effective mathematics is at predicting things we've never observed. Einstein's equations predicted the bending of light around massive objects before anyone had seen it happen. Dirac's mathematics predicted antimatter before it was discovered. The Higgs boson was predicted mathematically decades before particle accelerators could detect it.

Mathematics doesn't just describe what we observe—it predicts what we haven't seen yet, with precision that borders on the uncanny. This suggests something deeper: that reality doesn't just follow mathematical patterns, but might actually be mathematical in nature. Not described by math, but made of it.

If the universe is fundamentally computational—if physical laws are literally algorithms running on some substrate—then the "unreasonable effectiveness of mathematics" becomes perfectly reasonable. Math works because the universe is math. Physical laws are elegant equations because they are code, and good code tends toward elegance and efficiency.

This idea isn't new. Physicist John Archibald Wheeler proposed it in the 1980s with his famous phrase "it from bit"—the suggestion that physical reality emerges from information. Everything we think of as solid matter, from electrons to galaxies, might ultimately reduce to patterns of information being processed. Not described by information, but actually constructed from it at the most fundamental level.

Consider what this means. When you write a computer program, you create rules that govern how information changes over time. Variables have values. Functions transform those values according to specified operations. The program runs, and complex behavior emerges from simple rules applied consistently. The sophistication of modern software—operating systems, simulations, artificial intelligence—all emerges from binary operations performed trillions of times per second.

Now imagine the universe operating the same way. Physical laws are the rules. Particles are variables. Forces are functions that transform those variables according to mathematical operations. Time is the program running, executing these operations continuously. Reality emerges from information processing, just like the output of a computer program emerges from code execution.

This perspective makes several puzzling features of reality suddenly comprehensible. Take the speed of light, for instance. Nothing can exceed roughly 300,000 kilometers per second. This isn't a limitation of objects moving through space—it's a fundamental constraint built into spacetime itself. Why does this limit exist?

Standard physics explains it through special relativity: the speed of light falls out of the equations governing electromagnetic radiation and the structure of spacetime. But this just pushes the question back. Why do those equations have that form? Why that specific speed and not some other value?

In a computational universe, the speed of light might represent a processing limit—the maximum rate at which information can propagate through the simulation. Just as your computer has a maximum clock speed determining how fast it can process data, the universe might have a maximum speed determining how fast information can travel. The speed of light wouldn't be arbitrary. It would be a necessary feature of any reality running on finite computational resources.

Similarly, consider the Planck scale—the smallest meaningful length and time intervals in physics, roughly 10^-35 meters and 10^-43 seconds. Below these scales, space and time themselves become fuzzy and undefined. Quantum mechanics and general relativity both break down at the Planck scale, suggesting that spacetime is not infinitely divisible but has a granular structure.

This is exactly what you'd expect if reality were rendered at finite resolution. Digital images have pixels—discrete units below which you can't zoom further. Digital audio has sample rates—discrete time intervals that determine temporal resolution. A simulated universe would necessarily have some smallest unit of space and time, determined by the computational resources available to the simulators. That unit might be extraordinarily tiny, far below anything we can currently measure, but it would exist.

The Planck scale might be reality's pixel size—the resolution limit of the simulation. Not a metaphor. Literally the granularity of spacetime itself, imposed by computational constraints.

This interpretation gains support from quantum mechanics, which treats everything at the smallest scales as fundamentally discrete rather than continuous. Energy comes in quanta—discrete packets rather than smooth gradients. Angular momentum is quantized. Even properties like electric charge come in discrete units. The universe doesn't do smooth analog at the fundamental level. It does discrete digital, which is precisely what you'd expect from computational implementation.

Then there's the mystery of why physical laws take the specific mathematical forms they do. Why does gravity follow an inverse square law? Why do electromagnetic forces obey Maxwell's equations? Why does quantum mechanics use complex-valued wave functions?

These questions typically receive no answer beyond "that's how the universe works." We can describe the laws precisely, predict their consequences accurately, but we can't explain why these laws rather than infinitely many other possibilities. It's like asking why the English language uses the particular words it does—there's history and convention, but no deeper necessity.

Unless the laws are code. Then the question becomes: why did the programmers write these specific algorithms? And the answer might be: because these algorithms are efficient, elegant, and produce interesting emergent behavior. Good programmers value elegant solutions. Code that's unnecessarily complex or computationally expensive gets refactored. The algorithms running our universe might reflect the same engineering principles that human programmers use—simplicity, efficiency, modularity.

Consider gravity. The inverse square law means gravitational force decreases with the square of distance. Why? In a computational universe, this might be the most efficient way to implement attraction between objects in three-dimensional space. The same mathematical form governs electromagnetism, suggesting both might be different implementations of the same underlying computational principle.

Or consider conservation laws—energy conservation, momentum conservation, charge conservation. These principles state that certain quantities never change in total, only transform from one form to another. Why should the universe preserve these quantities so precisely? In computational terms, conservation laws might represent constraints imposed to maintain consistency and prevent errors. They're the universe's error-checking mechanisms, ensuring the simulation doesn't generate or destroy information improperly.

Even the apparent randomness of quantum mechanics might reflect computational efficiency. True randomness is expensive to compute—generating genuinely random numbers requires complex algorithms or external entropy sources. But simulations rarely need true randomness. They use pseudo-random number generators that produce statistically random-looking results from deterministic algorithms. The quantum randomness we observe might be deterministic at a deeper level, generated by some hidden algorithm we can't access from inside the simulation.

This brings us to a profound implication: if reality is computational, then physical laws aren't discovering eternal truths about existence. We're reverse-engineering someone else's code.

The comparison between universe and computer program becomes even more striking when we examine how reality handles complexity. Modern video games use a technique called "level of detail" rendering—objects far from the player are rendered at lower resolution, and details only load when the player gets close enough to see them. This saves enormous computational resources. Why render every brick in a distant building when the player can't see individual bricks at that distance?

Our universe appears to use the same strategy.

Quantum mechanics tells us that particles don't have definite properties until measured. An electron doesn't have a specific position and momentum simultaneously—it exists in a superposition of possibilities until observation forces it to "choose" definite values. This is deeply weird from a classical physics perspective. Why should observation matter? Why should reality care whether anyone's looking?

But it makes perfect sense from a computational perspective. Rendering every particle's exact state at every moment, everywhere in the universe, would require impossible computational resources. Far more efficient to leave things in probabilistic superposition—unrendered, in gaming terms—and only compute definite values when measurement occurs. Observation is when the simulation needs to render specific details because someone's looking.

This isn't just speculation. The mathematics of quantum mechanics explicitly describes unobserved particles as probability waves—distributions of potential values rather than definite states. Only measurement collapses the wave function, forcing reality to commit to specific values. It's as if the universe is saying: "I'll show you the details when you ask, but until then, I'm keeping things vague to save on processing power."

Wave-particle duality fits this pattern too. Light and matter behave as waves when unobserved, but as particles when measured. Why? Perhaps because waves are computationally cheaper to simulate. Instead of tracking billions of individual particles, you propagate wave equations that describe probabilistic distributions. Only when measurement requires particle-like behavior does the simulation instantiate actual particles with definite positions.

The double-slit experiment demonstrates this beautifully. Fire electrons one at a time through two slits, and they create an interference pattern over time—characteristic of waves passing through both slits simultaneously. But place a detector to see which slit each electron passes through, and the interference pattern disappears. The electrons behave as particles when watched, waves when not.

This has been experimentally verified thousands of times. Reality genuinely behaves differently depending on whether observation occurs. The standard interpretation is that measurement somehow "disturbs" the system, but this explanation has always felt unsatisfying. The disturbance would need to be retroactive, affecting the electron's behavior before it reached the detector, which violates causality.

The computational interpretation is cleaner: reality doesn't render details until required. Unobserved electrons propagate as probability waves because that's computationally efficient. Observed electrons manifest as particles with definite trajectories because that's what the measurement demands. The universe is optimizing its rendering strategy based on what information is actually needed.

Quantum entanglement presents another computational signature. When two particles are entangled, measuring one instantly affects the other, regardless of distance. Einstein called this "spooky action at a distance" and considered it evidence that quantum mechanics was incomplete. How could information travel faster than light between entangled particles?

But in a computational framework, no information needs to travel. Entangled particles are correlations in the simulation's database—linked variables that share computational resources. When you measure one, you're querying the database, which updates both entries simultaneously because they're logically connected at a level below spacetime. Distance is irrelevant because the correlation exists in the computational substrate, not in the rendered physical space.

It's like two characters in a video game whose health bars are linked by code. Damage one, and the other's health drops instantly, even if they're on opposite sides of the game world. No signal travels between them. The link exists in the game's code, operating at a level above the simulated space where distance matters.

Then there's quantum tunneling—particles appearing on the other side of energy barriers they shouldn't be able to cross classically. An electron faces a wall it doesn't have enough energy to climb over, yet sometimes appears on the other side anyway. Classical physics forbids this. Quantum mechanics predicts it through wave function mathematics, and experiments confirm it happens.

From a computational perspective, tunneling might represent the simulation occasionally violating its own classical rules when the probability calculations permit it. Classical physics is an approximation—useful rules that work most of the time but aren't fundamental. The deeper code allows for quantum effects that classical approximations don't capture. Tunneling is the simulation running its actual low-level algorithms instead of the classical shorthand.

Perhaps most telling is the delayed choice quantum eraser experiment, which suggests that measurements can affect behavior retroactively. Set up an experiment where you can choose, after a photon passes through a double-slit apparatus, whether to measure which slit it went through. The photon's behavior at the slits depends on whether you later measure its path—even though the photon passed the slits before you made that choice.

This appears to violate causality. How can a future decision affect past behavior? But in a computational universe, time might not work the way we intuitively assume. The simulation might not calculate events in strict temporal sequence. It might compute events as needed, rendering the past based on present measurements the same way video games render environments based on where the player looks. Causality is a feature of the rendered reality, not necessarily of the underlying computation.

All of this points toward the same conclusion: quantum mechanics' strange features—superposition, wave-particle duality, entanglement, tunneling, observer dependence—make more sense as computational optimizations and necessary features of a simulated reality than as mysterious properties of fundamental existence.

The mathematics of quantum mechanics even looks like code. The Schrödinger equation describes how quantum states evolve over time, and it's a differential equation that updates a system's state based on its current state—exactly how computer simulations work. Quantum operators manipulate quantum states through linear algebra operations. The entire formalism of quantum mechanics resembles information processing more than it resembles classical physics' continuous, deterministic evolution.

Physicists often note that quantum mechanics is fundamentally about information. The wave function encodes information about probabilities. Measurement extracts information from quantum systems. Entanglement is correlation of information. Quantum computation manipulates information in ways classical computers cannot. At its core, quantum mechanics is information theory applied to physical systems.

Or perhaps it's the other way around. Perhaps information isn't just a useful abstraction for understanding quantum mechanics. Perhaps information is the fundamental reality, and quantum mechanics is the low-level programming language of the universe—the actual algorithms running on whatever computational substrate implements our reality.

This perspective resolves many of quantum mechanics' philosophical puzzles. Why is there fundamental randomness? Because that's computationally efficient. Why does observation matter? Because that's when rendering occurs. Why do particles behave as probability waves? Because that's how the code represents uncomputed possibilities. Why is everything quantized and discrete? Because digital computation works with discrete values.

The universe increasingly looks less like a mechanical clockwork of particles colliding in space, and more like an elegant program processing information according to efficient algorithms. The "unreasonable effectiveness of mathematics" becomes the expected effectiveness of code. Physical laws aren't describing external reality—they're the syntax of the programming language in which reality is written.

And if reality is code, then somewhere, on some substrate we can't access, that code is running. The computational universe isn't just a metaphor. It might be a literal description of what existence is.

Chapter Three

Quantum Mechanics as Simulation Evidence

If you wanted to design a universe that was clearly simulated, you couldn't do much better than quantum mechanics. Its behavior is so counterintuitive, so divorced from everyday experience, so perfectly aligned with computational efficiency, that it almost seems designed to hint at its artificial nature, if you know what to look for.

The core strangeness of quantum mechanics revolves around measurement. Before you measure a quantum system, it exists in multiple contradictory states simultaneously. An electron is spinning both clockwise and counterclockwise. A photon is polarized both horizontally and vertically. A particle is in multiple locations at once. These aren't just unknown values waiting to be discovered. The mathematics and experiments both confirm that particles genuinely exist in multiple states until the moment of measurement.

Then measurement happens, and something changes. The superposition collapses. The particle "chooses" one definite state. The electron is spinning clockwise, not counterclockwise. The photon is horizontally polarized, not vertical. The particle is here, not there.

This is called the measurement problem, and it has troubled physicists for nearly a century. What counts as measurement? Why does measurement destroy superposition? What mechanism causes collapse? Is collapse physical or just a change in our knowledge? Different interpretations of quantum mechanics give different answers, but none are fully satisfying.

The Copenhagen interpretation, historically dominant, essentially says: measurement causes collapse, don't ask how, just calculate the probabilities and move on. This pragmatic approach produces accurate predictions but offers no deeper understanding. It's physics as recipe rather than explanation.

The many-worlds interpretation, discussed in Part One, proposes that measurement doesn't cause collapse—instead, reality branches into separate universes for each possible outcome. Superposition never ends; it just becomes multiple coexisting realities. This solves the measurement problem by denying that collapse happens, but it creates its own conceptual difficulties around the metaphysics of infinite splitting universes.

The simulation interpretation offers a third option: measurement is when the simulation computes and renders a definite value. Before measurement, the system exists in superposition because the simulation hasn't calculated specific values yet—only probability distributions. Measurement is the trigger that forces computation. The system doesn't "choose" a value randomly at measurement. The simulation computes a value based on its underlying algorithms, which appear random to us because we can't access the deeper computational layer.

Think about how a video game handles objects outside the player's view. Modern games don't fully render what you're not seeing. They maintain minimal data—positional information, basic states—but skip expensive rendering calculations for lighting, textures, and detailed physics. When you turn to look at something, the game rapidly computes and displays those details. From the player's perspective, everything seems to exist continuously, but the game is actually optimizing by only fully computing what's currently observed.

Quantum superposition looks remarkably similar. Particles maintain minimal information—probability distributions—until observation requires definite values. Then the simulation computes those values and renders them into existence. The collapse of the wave function isn't mysterious quantum magic. It's a computational operation, triggered by measurement, that converts probability distributions into specific outcomes.

This interpretation makes testable predictions. If superposition is computational optimization, then maintaining superposition should be "cheaper" than forcing definite states. Indeed, quantum coherence—keeping systems in superposition—is fragile. Any interaction with the environment causes decoherence, collapsing superposition into classical definite states. The universe seems eager to collapse superpositions, as if maintaining them requires effort or resources.

Similarly, measuring a quantum system becomes progressively harder as system size increases. Small systems like individual electrons remain in superposition easily. Large systems like cats or humans never show quantum behavior—they're always in definite classical states. This size dependence makes sense if computation cost scales with system complexity. Small systems are cheap to keep in superposition.

Large systems require so much computational overhead that the simulation just maintains them in classical definite states continuously.

The double-slit experiment reveals another computational signature. Fire particles one at a time through two slits, and they interfere with themselves as if each particle passes through both slits simultaneously. This interference creates a characteristic pattern on a detection screen—bands of high and low probability that could only arise from wave-like behavior.

But place detectors at the slits to measure which path each particle takes, and the interference vanishes. The particles behave as if they passed through one slit or the other, not both. The detection pattern changes from interference bands to simple accumulation—what you'd expect from classical particles taking definite paths.

The standard interpretation struggles with this. Why does placing a detector change particle behavior? The detector doesn't physically block or deflect the particles. It just observes them. Yet observation alone transforms wave-like behavior into particle-like behavior.

The computational interpretation is straightforward: without path detection, the simulation runs wave equations through both slits—computationally efficient. With path detection, the simulation must compute specific paths for each particle—more expensive, but required by the measurement. The universe renders what the experimental setup demands. No path measurement means wave propagation (cheap). Path measurement means particle trajectories (expensive but necessary).

This isn't observer consciousness causing magical effects. It's measurement apparatus demanding specific information, forcing the simulation to compute that information in response. The detectors are part of the simulation, not external observers mystically affecting reality with their minds.

Wave-particle duality generalizes this principle. Quantum entities aren't really waves or really particles. They're computational objects that manifest wave-like or particle-like properties depending on what information is extracted from them. Ask wave questions (interference, diffraction), get wave answers. Ask particle questions (position, momentum), get particle answers. The simulation provides the information you request in the form your measurement demands.

This is profoundly different from classical physics, where objects have definite properties independent of measurement. A classical ball has a specific position and momentum whether or not you measure them. Measurement reveals preexisting facts. But quantum measurement creates facts, computing them on demand rather than revealing what already existed.

Complementarity—Niels Bohr's principle that certain properties cannot be measured simultaneously—fits this framework perfectly. You can measure an electron's position or its momentum, but measuring one precludes measuring the other precisely. The uncertainty principle quantifies this limitation: the product of position uncertainty and momentum uncertainty must exceed a minimum value.

Why this restriction? In classical physics, no such limitation exists. You could, in principle, know both position and momentum with arbitrary precision. But quantum mechanics forbids it, not as practical difficulty but as fundamental law.

From a computational perspective, this makes sense as an anti-aliasing mechanism—a way to prevent excessive precision demands that would break the simulation. If observers could measure position and momentum simultaneously with infinite precision, they could reconstruct the wave function completely, extracting more information than the simulation can provide without revealing its discrete computational nature. The uncertainty principle might be enforced to maintain consistency and prevent information extraction beyond what the simulation can support.

Heisenberg originally derived the uncertainty principle from wave mathematics, showing it as a necessary consequence of representing particles as waves. But this just pushes the question back—why are particles represented as waves? Perhaps because waves are the natural representation for uncomputed possibilities in a simulation that computes definite values only when forced to by measurement.

The Copenhagen interpretation treated this operationally—"shut up and calculate." The many-worlds interpretation treated it metaphysically—infinite branching realities. The simulation interpretation treats it computationally—efficient rendering of an information-based reality.

All three produce the same experimental predictions. But only the simulation interpretation explains why quantum mechanics looks so much like optimized code.

Quantum entanglement pushes the computational interpretation even further. When two particles become entangled, measuring one instantly determines the state of the other, regardless of the distance separating them. Einstein famously objected to this "spooky action at a distance," arguing it violated the principle that nothing can travel faster than light. He believed quantum mechanics must be incomplete—that particles had hidden properties determining their behavior in advance, making the apparent instantaneous correlation an illusion.

Einstein was wrong. Experiments have repeatedly confirmed that entanglement is real, that no hidden variables explain the correlations, and that measuring one particle genuinely does affect its entangled partner instantly. The correlation happens faster than light could travel between them, yet no information is transmitted in a way that violates relativity. It's one of the most experimentally verified and conceptually baffling aspects of quantum mechanics.

How can this be? How do entangled particles "communicate" across vast distances without any signal passing between them?

They don't communicate. That's the wrong way to think about it.

In a computational universe, entangled particles are linked database entries—variables that share a constraint in the simulation's code. When you measure one, you're not sending a signal to the other. You're querying the simulation's state, and the simulation returns correlated values for both particles because that's how they were initialized. The correlation exists in the computational substrate, not in physical space. Distance is irrelevant because the link operates at a level where spacetime hasn't been rendered yet.

Imagine a multiplayer video game where two players' health bars are linked by code—damage to one reduces the other's health identically. The players could be on opposite sides of the game world, but their health values update simultaneously because the link exists in the game's backend code, not in the simulated game space. No signal travels through the game world. The correlation is enforced at a deeper level.

Quantum entanglement works the same way. The particles aren't separated in the simulation's computational space—they're linked variables, even if their rendered positions are light-years apart in spacetime. Measuring one particle forces the simulation to compute definite values for both, maintaining the correlation programmed when they were entangled. The apparent faster-than-light influence is an artifact of confusing the rendered spacetime with the underlying computational structure.

This resolves Einstein's paradox. Relativity says no information travels faster than light through space. And it doesn't. The correlation exists outside of space, in the simulation layer below spacetime. Relativity governs the rendered reality—spacetime and matter and energy. It doesn't govern the computational substrate implementing that reality.

Quantum tunneling provides another clue that reality operates more like probability code than classical mechanics. Place a particle against an energy barrier—a potential wall it shouldn't be able to cross based on its energy. Classically, it's like rolling a ball toward a hill: if the ball lacks enough energy to reach the top, it rolls back down. The ball never appears on the other side of the hill.

But quantum particles sometimes do appear on the other side. They "tunnel" through barriers they shouldn't be able to cross. The probability is often low, but it's not zero, and experiments consistently detect tunneling events in exact accordance with quantum mechanical predictions.

How does this happen? The wave function extends through the barrier, meaning there's nonzero probability of finding the particle beyond it. When measurement occurs on the far side, the simulation sometimes computes the particle as having tunneled through. It's not that the particle took a path through the barrier—quantum mechanics explicitly denies definite paths for unmeasured particles. Rather, the simulation allows probabilistic violations of classical energy conservation on quantum scales.

Why would a simulation permit this? Perhaps because strict classical rules are approximations—useful for macro-scale efficiency but not fundamental. The deep code allows quantum effects that classical physics doesn't capture. Tunneling is the simulation running its actual low-level algorithms instead of classical shortcuts.

This explains why tunneling probability decreases exponentially with barrier width and height. Thicker or taller barriers are "further" from classical expectations, requiring more extreme probability allowances. The simulation permits small violations readily, large violations rarely. It's like a video game allowing minor physics glitches that don't break immersion, but preventing major violations that would expose the simulation's artificial nature.

The delayed choice quantum eraser experiment might be the most simulation-like phenomenon in all of quantum mechanics. It suggests that measurements can retroactively determine behavior—that decisions made after a particle has been detected can change what the particle "did" before detection.

Here's how it works: photons pass through a double-slit apparatus and hit a detector. But the experimental setup includes additional optical elements that either preserve or destroy "which-path" information—data about which slit each photon traversed. Crucially, you can choose whether to preserve or destroy this information after the photon has already hit the detector.

If you choose to destroy which-path information, the detected photons show interference patterns—wave-like behavior suggesting they passed through both slits. If you choose to preserve which-path information, they show no interference—particle-like behavior suggesting they took definite paths. Your choice, made after detection, determines what pattern appears in the data from particles that hit the detector before you made that choice.

This seems to violate causality—future choices affecting past behavior. Physicists emphasize that no information travels backward in time, which is technically true. You can't use this effect to send messages to the past. But subjectively, it certainly feels like your measurement choice retroactively determines particle behavior.

The simulation interpretation handles this cleanly: the simulation doesn't necessarily compute events in strict temporal order.

It computes values as needed to maintain consistency. When you make a measurement choice, the simulation ensures all correlated events—including those earlier in time—are consistent with that choice. From inside the simulation, this appears as retroactive determination. From the computational layer, it's simply calculating a self-consistent set of values across multiple time points.

Video games do this routinely. They don't simulate every particle of smoke or every droplet of water in real-time. They compute visual effects when rendered, ensuring they're consistent with game state even if that state was determined "later" in real processing time. The illusion of continuous causal flow is maintained through careful computation, not through calculating everything in strict temporal sequence.

Reality might work the same way. Time as we experience it—one moment flowing to the next—might be a feature of the rendered simulation, not the underlying computation. The simulation ensures self-consistency across time, computing values as needed rather than in rigid temporal order. Delayed choice experiments reveal this flexibility by creating situations where temporal ordering matters to our intuitions but not to the simulation's actual operation.

All of these quantum phenomena—superposition, measurement collapse, wave-particle duality, entanglement, tunneling, delayed choice effects—share a common feature. They violate our classical intuitions about reality while perfectly matching what you'd expect from an efficiently coded simulation. They're exactly the behaviors you'd build into a universe if you wanted to minimize computational overhead while maintaining consistent behavior across all possible measurements.

Quantum mechanics isn't evidence that reality is weird. It's evidence that reality is code.

Chapter Four

The Fine-Tuning Problem

The universe shouldn't exist. Or more precisely, the universe shouldn't exist in a form capable of supporting complexity, structure, or life. The odds against it are so astronomical that they strain comprehension.

This isn't poetic exaggeration. It's mathematical fact. The fundamental constants of physics—the numbers that determine how forces operate, how particles interact, how matter behaves—are calibrated to levels of precision that make winning the lottery look like a certainty. Change any of these constants by even tiny amounts, and the universe becomes sterile: no atoms, no stars, no planets, no possibility of complexity or consciousness.

This is the fine-tuning problem, and it's one of the most profound mysteries in physics. Why do the fundamental constants have the exact values they do? And why are those values so precisely calibrated to permit our existence?

Start with the cosmological constant—the energy density of empty space itself. This value determines whether the universe expands, contracts, or remains stable. Quantum field theory predicts that empty space should be seething with virtual particles, contributing enormous energy density. Calculate what that density should be based on particle physics, and you get a number roughly 10^120 times larger than what we observe.

This is not a small discrepancy. It's the worst prediction in the history of physics—off by 120 orders of magnitude. Something cancels out almost all of that predicted vacuum energy, leaving a tiny residual value that's just right for the universe to expand at a rate that allows structure formation.

If the cosmological constant were much larger, space would expand so rapidly that matter could never clump together to form galaxies, stars, or planets. If it were much smaller—or negative—the universe would have collapsed back on itself shortly after the Big Bang. The actual value sits in an incredibly narrow range that permits the gradual expansion and structure formation we observe.

No one knows why. The standard model of particle physics offers no mechanism for why the cosmological constant should have this value. It's simply a number we measure and plug into our equations. But that number appears to be calibrated to one part in 10^120 to make a universe like ours possible.

Then there's the strong nuclear force—the interaction that binds quarks into protons and neutrons, and binds those nucleons into atomic nuclei. Its strength is characterized by a constant that determines how tightly nuclei hold together. Increase that constant by just two percent, and deuterium—an isotope of hydrogen with one proton and one neutron—becomes so stable that all hydrogen in the early universe would have fused into deuterium and heavier elements. No hydrogen means no water, and stars would burn out far too quickly for life to develop.

Decrease the strong force by five percent, and deuterium becomes so unstable that it can't form at all. Without deuterium as an intermediate step, you can't build heavier elements. The universe would contain only hydrogen—no carbon, no oxygen, no chemistry beyond the simplest atom. No complexity of any kind.

The electromagnetic force presents similar precision requirements. It governs how charged particles interact, determining the strength of chemical bonds and the structure of atoms. Make it slightly stronger, and electrons bind so tightly to nuclei that chemical reactions become impossible. Make it slightly weaker, and atoms become so large and fragile that stable molecules can't form. Either way, chemistry—and thus biology—becomes impossible.

The electron's mass relative to the proton's mass is another critical parameter. If electrons were slightly heavier, atoms would be smaller and chemical bonds would break down. If electrons were slightly lighter, atoms would be too large and unstable. The actual mass ratio is about 1/1836—a seemingly arbitrary number that falls in the narrow range permitting stable chemistry.

The ratio of the electromagnetic force to the gravitational force is even more striking. Gravity is extraordinarily weak compared to electromagnetism—about 10^36 times weaker. This vast difference seems strange, but it's essential. If gravity were much stronger relative to electromagnetism, stars would burn out in thousands of years rather than billions, giving insufficient time for life to evolve. If gravity were much weaker, stars would never ignite at all—gravity wouldn't compress matter enough to trigger fusion.

The expansion rate of the early universe represents another knife-edge balance. Too fast, and matter would have spread too thin for gravity to pull it into galaxies and stars. Too slow, and the universe would have collapsed back on itself. The actual expansion rate at one second after the Big Bang had to be accurate to one part in 10^15 to produce a universe remotely like ours.

The entropy of the early universe—its degree of order versus disorder—required similar precision. Low entropy means high order. The Big Bang started in an extraordinarily low entropy state, which seems unlikely because high entropy states vastly outnumber low entropy states. Yet the universe began in a highly ordered configuration, permitting the gradual increase of entropy that drives time's arrow and enables complex structures to form temporarily before eventual heat death.

Calculate the probability of the universe randomly starting in such a low entropy state. Roger Penrose, one of the world's leading physicists, estimated it as one part in 10^10^123. That's not 10^123. That's ten to the power of ten to the power of 123—a number so large that writing it out would require more digits than there are particles in the observable universe.

Carbon formation provides a particularly striking example of fine-tuning. Carbon is essential for life as we know it, but creating carbon inside stars requires a seemingly miraculous coincidence. Carbon forms when two helium nuclei fuse to create beryllium, which then captures another helium nucleus to become carbon. But beryllium is highly unstable—it decays in 10^-16 seconds. For carbon to form in any quantity, the energy levels of helium, beryllium, and carbon must align in a resonance that dramatically increases the reaction rate.

Physicist Fred Hoyle predicted this resonance must exist, purely from the fact that carbon exists in our universe. When experimentalists looked, they found it exactly where Hoyle predicted, at an energy level that's "just right" for producing the carbon abundance we observe. Had that energy level been even slightly different—four percent higher or lower—the universe would contain essentially no carbon, and carbon-based life would be impossible.

Hoyle, an atheist, later wrote that this apparent coincidence "looks like a put-up job"—as if the universe were designed specifically to produce carbon and enable life. The alternative explanations—pure chance or some deeper principle we don't understand—both seem inadequate when confronted with the precision involved.

The list continues. The ratio of matter to antimatter in the early universe. The number of spatial dimensions. The relative strengths of the four fundamental forces. The Higgs field strength determining particle masses. Every one of these parameters falls within a narrow range that permits complexity and structure. Step outside those ranges, and you get sterile universes: empty space expanding forever, or collapsing immediately, or filled with only the simplest particles unable to form atoms, let alone stars or life.

How narrow are these ranges? Cosmologist Luke Barnes has calculated that if you randomly vary the fundamental constants, fewer than one in 10^136 parameter combinations would produce a universe with stars. That's being generous—it only requires stars, not planets or chemistry or anything else necessary for life. The probability of randomly getting all the conditions right for life is far lower.

This is the fine-tuning problem. The universe appears exquisitely calibrated to produce observers like us. Why?

Three major explanations compete to explain fine-tuning: luck, necessity, or design. Each has profound implications.

The luck hypothesis says the constants just happen to have the values they do. We won the cosmic lottery. No deeper explanation exists—the universe rolled the dice and came up with parameters compatible with our existence. This is logically possible but deeply unsatisfying. Invoking luck for one or two coincidences might be reasonable. But invoking luck for dozens of independent parameters, each calibrated to extraordinary precision, strains credibility.

It's like shuffling a deck of cards and getting perfect numerical order every time for a hundred shuffles. Technically possible, but so improbable that you'd reasonably suspect the deck was rigged. The fine-tuning of cosmic constants is far more extreme than sorted cards—more like shuffling atoms randomly and getting a functioning computer.

The necessity hypothesis proposes that the constants must have their observed values due to some deeper physical law we haven't discovered yet. Perhaps there's an underlying principle that constrains the constants, making our universe the only possibility or one of very few possibilities. This would be elegant—physics traditionally seeks necessary relationships, not arbitrary coincidences.

But no one has found such a principle. Every attempt to derive the fundamental constants from deeper laws has failed. String theory once promised to explain why the constants have their values, but it turned out to predict a "landscape" of 10^500 possible universes with different constants rather than singling out our specific values. The necessity hypothesis remains purely speculative, with no evidence supporting it and substantial theoretical obstacles preventing it.

The design hypothesis suggests intentionality—that someone or something set the parameters deliberately to produce a life-permitting universe. Theists interpret this as God creating the universe for purpose. But design doesn't require supernatural deities. It could mean our universe is simulated, with programmers choosing initial conditions and physical laws to achieve desired outcomes.

This explanation fits the evidence remarkably well. Programmers setting up a simulation would choose parameters to achieve specific goals—perhaps studying how complexity emerges, or running ancestor simulations of evolutionary history, or simply creating interesting environments. They'd tune the constants carefully, just as video game designers tune their game physics to produce enjoyable gameplay. Fine-tuning becomes expected rather than miraculous.

Critics often invoke Occam's Razor against design explanations. Why multiply entities unnecessarily? Why propose simulators or gods when simpler explanations might exist? But this objection has less force than it initially appears. The question isn't whether design is simpler in absolute terms—it's whether it's simpler than the alternatives for explaining the data we observe.

If the universe required no fine-tuning—if most parameter combinations produced habitable universes—then design would be unnecessarily complex. But we don't live in that universe. We live in one where dozens of parameters need extreme precision, and where no known physical principle explains that precision. Given this data, design might actually be the simpler explanation compared to imagining an undetected deeper law or accepting astronomically improbable luck.

The multiverse hypothesis offers a middle path between luck and design. It proposes that countless universes exist with different values for their fundamental constants. Most are sterile and lifeless. But with enough universes—perhaps infinitely many—some will randomly have life-permitting constants. We necessarily find ourselves in one of those rare habitable universes because we couldn't exist in the sterile ones.

This is the anthropic principle applied cosmologically. Just as we necessarily find ourselves on a planet with liquid water and breathable atmosphere—not because Earth is specially designed, but because we couldn't observe ourselves on an uninhabitable planet—we necessarily find ourselves in a life-permitting universe, even if such universes are rare.

The multiverse hypothesis has gained traction in physics, particularly through eternal inflation theory. This proposes that quantum fluctuations drive eternal exponential expansion in some regions while other regions stop inflating and form bubble universes like ours. Each bubble has randomly determined physical constants. If eternal inflation is correct, the multiverse is real, and fine-tuning is explained through anthropic selection.

But the multiverse has significant problems as an explanation. First, it's scientifically problematic because other universes are potentially unobservable in principle. If they can't affect our universe and we can't measure them, in what sense do they exist? The multiverse risks becoming unfalsifiable metaphysics rather than testable science.

Second, even if multiverses exist, that doesn't eliminate design—it just pushes it back a level. Who or what set up the multiverse-generating mechanism? Why does eternal inflation produce universes with varied constants rather than no universes at all? The meta-laws governing multiverse generation might themselves require fine-tuning. You've replaced one fine-tuning problem with another.

Third, the multiverse doesn't explain why our universe has additional features that aren't strictly necessary for life. The universe appears mathematically elegant, with physical laws taking beautiful symmetric forms. Conservation laws, gauge symmetries, the connection between symmetries and forces—these suggest intentional design choices rather than random parameter selection. A randomly selected life-permitting universe from a vast multiverse should look barely habitable, with constants just barely inside the ranges that allow life. Our universe seems better tuned than necessary—optimized rather than minimally adequate.

The simulation hypothesis naturally explains this optimization. Programmers wouldn't just make a universe that barely permits life. They'd optimize it—choosing elegant mathematical structures, implementing efficient computational rules, creating a universe that's not only functional but beautiful in its underlying principles. The elegance of physical law suggests intentional design choices, not random selection from a multiverse.

Fine-tuning also exhibits a suspicious property: the parameters are tuned not just for any complexity, but specifically for the kind of complexity that produces observers. Chemistry works. Stable atoms form. Stars have long lifetimes. Planets orbit in stable configurations. The universe permits information processing and consciousness to emerge through evolutionary processes.

If you were designing a simulation to study consciousness or run ancestor simulations or create interesting emergent behavior, these are exactly the features you'd want. You'd tune the constants to make observers likely, even inevitable. The fact that our universe is observer-friendly rather than just structure-friendly suggests purposeful configuration.

Consider an analogy. You wake up in a room perfectly calibrated for human survival—air pressure at one atmosphere, temperature at twenty degrees Celsius, oxygen at twenty-one percent, gravity at 9.8 meters per second squared. Do you conclude: (1) pure luck that you randomly appeared in the narrow parameter range supporting human life, (2) physical necessity that rooms must have these properties, or (3) someone designed this room for human habitation?

Obviously option three. And our universe is that room, calibrated to far greater precision across far more parameters. The simulation hypothesis simply applies the same reasoning at cosmic scale.

This doesn't prove we're simulated. Fine-tuning could have other explanations we haven't thought of. But it's striking that the one explanation humans intuitively accept for comparable fine-tuning in human-made artifacts—intentional design—is the explanation many people resist most strongly when applied to the universe itself.

Perhaps that resistance reflects discomfort with the implications. If the universe is designed, that raises questions about the designers—their nature, their motives, their relationship to us. These questions make many people uncomfortable because they can't be answered empirically, at least not with current methods.

But comfort isn't a guide to truth. And fine-tuning is real, measured, quantified. The parameters are calibrated to precision that defies luck and has no known necessary explanation. Something set those values. Either reality rolled impossible dice, or some principle we don't understand constrains the constants, or someone configured the initial conditions deliberately.

The simulation hypothesis takes that third option seriously. It proposes that fine-tuning reflects intentional parameter selection by whoever initialized our reality. The constants are calibrated because programmers calibrated them, the same way game designers tune physics engines to produce desired gameplay.

If we're simulated, fine-tuning stops being miraculous. It becomes expected—the natural consequence of intentional design choosing initial conditions to achieve specific goals. The mystery isn't why the universe is fine-tuned. The mystery is what goals the fine-tuning was meant to achieve.

Chapter Five

Glitches in the Matrix

If we live in a simulation, we should expect occasional glitches—moments when the program fails to maintain perfect consistency, when rendering errors occur, when the illusion of continuous reality briefly falters. These wouldn't necessarily be obvious bugs. A well-designed simulation would handle most errors gracefully, correcting inconsistencies before observers notice them. But perfect consistency is computationally expensive, and even the best programs have edge cases where unexpected behavior emerges.

Humans report experiencing such anomalies constantly. Déjà vu. The Mandela Effect. Impossible coincidences. Moments when reality seems to hiccup or repeat. Skeptics dismiss these as quirks of human psychology—memory errors, pattern recognition gone awry, confirmation bias. And certainly, human perception is flawed. Our brains create illusions, false memories, and spurious connections regularly.

But what if some of these experiences aren't entirely psychological? What if they're occasional glimpses of the computational nature underlying reality—brief moments when the simulation's mechanics become visible to those inside it?

Déjà vu is the most common reported glitch. You enter a room and feel certain you've experienced this exact moment before, down to specific details. The people, the conversation, the lighting—everything feels impossibly familiar, as if you're rewatching a scene you've already seen. The sensation typically lasts seconds, then fades, leaving you disoriented and uncertain whether it was memory or something stranger.

Neurologists have studied déjà vu extensively. They've identified brain regions involved—particularly the temporal lobe and hippocampus, areas crucial for memory formation. They've documented that certain epilepsy patients experience déjà vu frequently, suggesting it results from abnormal neural firing. They've shown that déjà vu becomes more common under stress, fatigue, or certain medications. All of this suggests a biological explanation: misfiring neurons create false signals of familiarity when processing novel experiences.

But there's another interpretation. In a simulated reality, memory isn't stored only in brains—it's data in the simulation's databases. If the simulation occasionally repeats a scene—perhaps loading a cached segment, or correcting an error by rolling back and replaying events—observers might experience genuine repetition of moments. The simulation runs a segment, detects an error, resets, and runs it again. From the observer's perspective, this manifests as déjà vu—the eerie sensation that you've lived this exact moment before, because you have, in a previous iteration that was subsequently erased and rerun.

This would explain why déjà vu feels qualitatively different from normal memory confusion. Regular false memories don't come with that uncanny certainty that you're re-experiencing something. They're vague, questionable, easy to doubt. Déjà vu has an immediate, visceral quality—you feel absolutely certain of the repetition, even though you know rationally it shouldn't be possible. Perhaps that certainty reflects genuine repetition, with your consciousness briefly retaining traces of the previous iteration even after the simulation reset local memory.

The Mandela Effect represents a more complex category of potential glitches. It refers to collective false memories—situations where large groups of people remember historical events or details differently from documented reality. The name comes from many people claiming to remember Nelson Mandela dying in prison in the 1980s, when he actually died in 2013 after serving as South Africa's president. Other common examples include people remembering "Berenstain Bears" as "Berenstein Bears," or recalling a 1990s movie called "Shazaam" starring comedian Sinbad as a genie—a movie that apparently never existed.

Psychologists explain the Mandela Effect through social contagion of false memories. One person's confusion spreads through communities, particularly online where people reinforce each other's false recollections. Misremembering "Berenstain" as "Berenstein" is natural—the latter spelling is more common and easier to pronounce. Confusing one genie movie (Shaquille O'Neal's "Kazaam") with another that never existed is simple memory contamination. Nothing mysterious required.

Yet the consistency and specificity of some Mandela Effects is intriguing. Large numbers of people remember not just vague details but specific scenes, quotes, or images that differ from documented reality. These memories often include sensory details and emotional content that seems too vivid to be entirely fabricated. Could this reflect timeline edits—the simulation retroactively changing events and updating documentation, but failing to perfectly synchronize all memories?

In a simulated reality, the past isn't fixed. It's data that can be modified. If the simulation needs to correct an error or change a narrative, it could edit history directly, updating all documentation and most memories to reflect the new timeline. But perhaps memory erasure isn't perfect—some minds retain fragments of the original timeline, creating conflict between what they remember and what the evidence now shows. The Mandela Effect would represent failed memory synchronization after timeline edits.

This interpretation is highly speculative and unfalsifiable, which makes it scientifically problematic. You can't prove the simulation edited history if all documentation supports the current version and only memories contradict it. Memories are notoriously unreliable, making them weak evidence against documented facts. Still, the subjective experience of remembering something clearly, having multiple people share that memory, then finding unanimous documentation showing something else—that's an odd pattern worth noting, even if mundane explanations suffice.

Synchronicity—meaningful coincidences that seem too perfect to be random—presents another category of potential glitches. Carl Jung coined the term to describe acausal connections: events related by meaning rather than cause and effect. You think of an old friend you haven't contacted in years, and they call moments later. You dream about a specific scenario, and it plays out the next day. You need information, and a book falls off a shelf opened to the exact page containing what you need.

Statistically, we should expect some coincidences. With billions of people having thousands of experiences daily, unlikely events become likely in aggregate. Someone will win the lottery, even though any specific person probably won't. Most coincidences are mundane chance—the universe contains enough random events that some will align meaningfully by pure probability.

But synchronicity advocates argue that some coincidences exceed statistical expectation, particularly when they cluster around specific people or situations. They note that coincidences often serve functional purposes—providing needed information, preventing accidents, facilitating unlikely meetings that prove important. Pure chance shouldn't care about function, yet synchronicities frequently seem helpful rather than random.

In simulation terms, synchronicities might represent direct intervention—the simulation's operators adjusting probabilities to guide narratives toward desired outcomes. If the simulation serves a purpose—studying consciousness, running ancestor simulations, generating interesting stories—then occasional probability manipulation might be necessary to keep things on track. Most events run on normal physics and probability. But at critical junctures, the simulation tweaks odds to produce outcomes that serve its purpose.

This would explain why synchronicities often feel meaningful rather than random. They are meaningful—they're intentional adjustments made for reasons the simulation's operators understand, even if we don't. The coincidence that saves your life might be the simulation ensuring your consciousness continues, similar to quantum immortality's mechanism. The unlikely meeting that changes your life trajectory might be narrative engineering, pushing your story toward more interesting outcomes.

Again, this is unfalsifiable. Any coincidence, no matter how unlikely, can be explained as chance. And human brains are excellent at finding patterns and meaning in randomness. We're predisposed to see agency and intention, even when none exists. Believers in synchronicity might simply be overinterpreting normal statistical variation.

But the subjective experience remains striking. When you experience a coincidence so perfectly timed and specifically relevant that it feels like reality responded to your thoughts or needs, dismissing it as pure chance requires effort. It feels designed. And perhaps, occasionally, it is.

Physical anomalies present harder evidence of potential glitches—documented phenomena that violate expected behavior without clear explanation. These aren't subjective experiences or memory problems. They're measurable deviations from predicted outcomes that persist despite investigation.

Take the Pioneer anomaly. NASA's Pioneer 10 and 11 spacecraft, launched in the early 1970s to explore the outer solar system, exhibited unexpected deceleration. They slowed down slightly more than gravitational calculations predicted. The effect was tiny—about 8.5 × 10^-10 meters per second squared—but consistent and measurable over decades of tracking data.

For years, this remained unexplained. Scientists proposed various hypotheses: gravitational effects from undiscovered objects, interactions with interplanetary medium, modifications to gravitational theory itself. Eventually, researchers concluded that thermal radiation from the spacecraft's plutonium power sources created asymmetric thrust, accounting for most or all of the anomaly. A mundane explanation for an initially mysterious observation.

But what if some anomalies lack mundane explanations? What if occasional measurements deviate from predictions not because our models are incomplete, but because the simulation occasionally produces computational errors that manifest as physical law violations?

The flyby anomaly might be such a case. Multiple spacecraft performing Earth flybys—using the planet's gravity to change trajectory—have exhibited unexpected velocity changes. The effect is small and inconsistent: some flybys show anomalous acceleration, others don't, with no clear pattern. Various explanations involving Earth's rotation, atmospheric drag, and tidal effects have been proposed, but none fully account for all observations.

The inconsistency is what's intriguing from a simulation perspective. Consistent deviations suggest systematic errors in our models. Inconsistent deviations—some flybys affected, others not—suggest something more erratic. Perhaps computational precision varies depending on system load. Perhaps the simulation uses different calculation methods depending on circumstances, producing occasional rounding errors. Perhaps these are genuine glitches that occur when multiple complex calculations interact in unexpected ways.

The Fermi Paradox might represent the largest-scale potential glitch: the apparent absence of extraterrestrial civilizations despite high probability they should exist. The universe contains billions of galaxies, each with billions of stars, many with planets. Even conservative estimates suggest millions of potentially habitable worlds in our galaxy alone. Given the universe's age, some civilizations should have arisen millions or billions of years before humanity, achieving technological capabilities far beyond ours.

Where are they? Why haven't we detected signals, encountered probes, or found evidence of megastructure engineering? The silence is deafening and demands explanation.

Standard answers include: intelligent life is extremely rare, civilizations tend toward self-destruction, interstellar travel is impractical, or aliens exist but avoid contact. Each explanation has merit but also problems. Maybe life is rare—but how rare must it be to leave us apparently alone in billions of galaxies? Maybe civilizations self-destruct—but all of them, everywhere, without exception? Maybe interstellar travel is impossible—but why should that be a universal constraint?

The simulation hypothesis offers a different answer: maybe we're alone because the simulation only fully renders our local region. Other star systems exist as background—sufficient data to pass astronomical observations, but without the computational expense of simulating full civilizations. It's like video game skyboxes—distant scenery that looks three-dimensional but is actually flat images. Or like non-player characters in role-playing games who have basic behavioral scripts but not full consciousness.

If Earth and humanity are the focus of the simulation—perhaps we're the observed civilization in an ancestor simulation, or the interesting case study in a research project—then simulating billions of other conscious civilizations would be wasteful. The simulation renders our local environment in detail because we interact with it. Distant stars are low-detail background, sufficient for consistency but not requiring full physics simulation.

This would explain not just the absence of aliens but also why SETI searches have found nothing. There's nothing to find. The simulation generates radio noise consistent with natural astrophysical processes, but no artificial signals because no other civilizations are actually being simulated to generate them.

Critics correctly note this is unfalsifiable—you can't prove aliens don't exist elsewhere just because we haven't found them yet. The universe is vast, and our searches have been limited. Absence of evidence isn't evidence of absence. Fair enough. But the Fermi Paradox remains puzzling, and the simulation hypothesis offers a coherent explanation that standard astrophysics struggles to provide.

Near-death experiences might offer the most tantalizing hints of reality's edges—the boundaries where the simulation's rendered environment ends. People who come close to dying often report similar experiences: moving through tunnels toward light, encountering deceased relatives, experiencing overwhelming peace or joy, perceiving events from outside their bodies, then returning to normal consciousness.

Materialist neuroscience explains these as hallucinations produced by oxygen-deprived or dying brains. Certain neurochemicals—endorphins, DMT, serotonin—create altered states resembling NDE reports. Temporal lobe stimulation can produce out-of-body sensations. Everything about NDEs can be replicated through brain manipulation, suggesting they're purely neurological phenomena with no metaphysical significance.

But some NDE features are harder to explain neurologically. Patients under general anesthesia, with no measurable brain activity, sometimes report accurate perceptions of surgery and conversations—details they couldn't have known through normal sensory channels. Blind patients report visual experiences during NDEs, describing scenes later verified as accurate. These cases are controversial and disputed, but they're documented in medical literature and resist easy dismissal.

In simulation terms, NDEs might represent consciousness approaching the simulation's boundaries. When the body nears death, consciousness—the actual process being simulated—begins transitioning out of the simulation's rendered environment. The tunnel, light, and deceased relatives might be artifacts of that transition: loading screens, welcome interfaces, or the edge of the simulation appearing as metaphorical imagery because consciousness lacks better frameworks for interpreting raw computational transitions.

The accurate perceptions during zero brain activity are particularly suggestive. If consciousness is implemented in the simulation substrate rather than generated by brains, then brain cessation wouldn't immediately end consciousness—just its ability to interact with the simulated body. During the transition period, consciousness might access information through the simulation layer directly, bypassing normal sensory channels. Hence veridical perceptions without functional sensory organs or brain activity.

This is speculative, and skeptics reasonably point out that all NDE evidence is anecdotal or involves small samples with methodological difficulties. No large-scale, rigorously controlled study has definitively proven veridical perception during confirmed brain death. Most NDEs probably are neurological hallucinations. But if even a few cases represent genuine perception without brain function, that would be profound evidence that consciousness isn't generated by brains but exists at a deeper level—exactly what simulation theory predicts.

All these potential glitches share a pattern: they're at the boundary between clearly explainable and genuinely anomalous. Déjà vu is probably neurological, but feels too precise. The Mandela Effect is probably false memory, but shows odd consistency. Synchronicities are probably coincidence, but seem too meaningful. Physical anomalies are probably measurement error or incomplete models, but resist final explanation. The Fermi Paradox probably has a standard solution, but none fully convince. NDEs are probably hallucinations, but occasionally include verified information.

None of these individually prove we're simulated. All have mundane explanations that suffice under Occam's Razor. But collectively, they paint a picture of reality occasionally behaving strangely in ways consistent with computational implementation having edge cases, rendering limits, and occasional errors.

If we're not simulated, these are just quirks of human psychology and gaps in scientific knowledge. If we are simulated, they're glimpses of the machinery underlying reality—brief moments when the code becomes visible to those running inside it.

The question is whether pattern recognition is finding real patterns, or creating them from noise. And from inside the simulation, that question might be unanswerable.

Chapter Six

Consciousness and the Observer

Consciousness is the hardest problem in all of science. We can explain how brains process information, how neurons fire in response to stimuli, how sensory data gets transformed into behavioral outputs. We can map brain regions to functions, trace neural pathways, measure electrical activity with extraordinary precision. But none of this explains why there's something it's like to be you. Why subjective experience exists at all.

This is called the hard problem of consciousness, and it remains completely unsolved. The easy problems (explaining cognitive functions like memory, attention, learning) have made enormous progress. We understand mechanistically how brains perform these tasks. But the hard problem (explaining why performing these tasks feels like anything) has resisted every attempt at explanation.

Why should neurons firing create the sensation of seeing red? Why should information processing generate the experience of tasting coffee or feeling pain or hearing music? Physical processes in the brain correlate with conscious experiences, but correlation isn't explanation. We can predict that certain neural activity will produce certain experiences, but we can't explain why those physical processes should generate subjective experience at all rather than simply processing information unconsciously.

Philosopher David Chalmers, who formalized the hard problem, argues that no amount of information about physical brain processes could explain subjective experience. Even if we knew everything about how neurons work, every connection in the brain, every chemical interaction, we still wouldn't understand why it feels like something to be that system. The explanatory gap between physical processes and subjective experience appears unbridgeable.

This gap creates serious problems for materialist philosophies that claim consciousness emerges from physical processes. If consciousness were just neurons firing in complex patterns, we should be able to explain subjective experience by explaining those patterns. But we can't. And many philosophers and scientists suspect we never will, because subjective experience isn't a property that emerges from physical complexity. It's something fundamentally different from physical processes.

The simulation hypothesis offers a potential resolution. If consciousness isn't generated by brains but exists in the simulation substrate (if subjective experience is implemented at a computational level below the rendered physical reality), then the hard problem isn't solvable from inside the simulation because we're asking the wrong question. We're trying to explain consciousness as emerging from brains, when actually brains are the interface through which consciousness interacts with simulated bodies.

Think of it this way: video game characters don't generate their own behavior. The game code runs on external hardware, and characters are manifestations of that code within the game world. If you were a character trying to understand where your actions come from by studying your rendered body, you'd fail. Your behavior originates outside your visible form, in code running on hardware you can't perceive.

Similarly, if consciousness is implemented in the simulation substrate, studying brains won't reveal its origin. Brains would be input/output devices connecting substrate-level consciousness to simulated bodies. Damage the brain, and the connection degrades or severs. But the consciousness itself exists at a deeper level, just as your computer continues running even if you disconnect the monitor.

This explains why the hard problem seems intractable. We're looking for consciousness in the wrong place. We're examining the rendered interface and wondering why it generates subjective experience, when actually it doesn't. It merely manifests it. The experience originates elsewhere, in computational processes we can't directly observe.

But this raises an even deeper question: what is consciousness in computational terms? If it's implemented in the simulation substrate, what's being implemented? What kind of process could generate subjective experience?

Integrated Information Theory, proposed by neuroscientist Giulio Tononi, offers one approach. It suggests consciousness is a fundamental property of systems that integrate information. Systems with internal causal structure creating unified experience greater than the sum of their parts. The more integrated information a system possesses, the more conscious it is. A rock has near-zero integrated information. A brain has enormous integrated information, hence rich conscious experience.

This theory treats consciousness as intrinsic to information processing rather than emerging from it mysteriously. Wherever sufficiently integrated information processing occurs, consciousness exists.

It's not that complex systems coincidentally develop consciousness. Consciousness is what integrated information processing is, viewed from the inside.

If IIT is correct, then consciousness could be implemented computationally. A sufficiently complex simulation running algorithms that integrate information would generate genuine subjective experience. The substrate wouldn't matter. Silicon, neurons, or exotic computational hardware all work equally well. What matters is the pattern of information integration.

This makes simulation theory much more plausible. If consciousness requires only information integration, not biological neurons specifically, then simulated beings would be genuinely conscious. Their experiences would be real experiences, their suffering real suffering, their joy real joy. They wouldn't be philosophical zombies (entities that behave consciously but lack inner experience). They'd be conscious in the same fundamental way humans are, just implemented on different substrate.

But how do we know we're not philosophical zombies? How do you know other people are conscious rather than simply behaving as if they are? This is the problem of other minds, and it has no definitive solution. You know you're conscious through direct experience. You infer others are conscious because they behave like you and have similar brains. But you can't verify their subjective experience directly.

In a simulation, this problem intensifies. Are other people in your simulation genuinely conscious, or are they NPCs (non-player characters running behavioral scripts without inner experience)? If the simulation is trying to be realistic, it would implement some entities as fully conscious (probably including you, since you're experiencing this question). But it could economize by making background characters unconscious automatons.

This possibility is disturbing. The person you're talking to might be conscious, or might be an NPC executing a conversation script without actually experiencing anything. You have no way to tell. They act conscious, respond appropriately, claim to have inner experience. But so would a well-programmed NPC. The simulation could be solipsistic. You're the only conscious observer, and everything else is rendered background.

More likely, if the simulation serves a purpose like ancestor simulations or social research, many entities would be conscious. Creating billions of unconscious NPCs to interact with one conscious being would be inefficient. Most of the simulation's complexity would be wasted. More efficient to make many entities conscious, creating a population whose interactions and experiences can be studied.

But we might still expect a hierarchy. Entities directly relevant to the simulation's purpose would be fully conscious. Background characters (people you'll never interact with meaningfully) might be lower-resolution consciousnesses, or NPCs with behavioral scripts. The simulation would render consciousness at different levels of detail, just as it renders physics and sensory information at different resolutions depending on observer proximity.

This creates an uncomfortable question: how conscious are you? Are you a fully-detailed consciousness running at maximum resolution, or a lower-detail implementation sufficient for your narrative role? You feel fully conscious, but lower-resolution consciousness might also feel fully conscious to itself. It would lack the metacognitive capacity to recognize its own limitations.

Perhaps consciousness itself is variable. Not binary (conscious or unconscious) but a spectrum. Simple systems have minimal consciousness. Complex systems have rich consciousness. You might be more or less conscious than you assume, with no internal way to determine where you fall on the spectrum.

The binding problem offers another clue about consciousness's relationship to simulation. Your brain doesn't process information in one place. Visual processing happens in the occipital lobe. Auditory processing in the temporal lobe. Language in frontal and temporal regions. Memory formation in the hippocampus. Motor control in the motor cortex. Dozens of specialized regions operate in parallel, each handling specific functions.

Yet you experience unified consciousness. A single coherent perspective integrating all these disparate processes. You don't experience vision separately from hearing separately from thought. You experience them together, bound into one seamless whole. How does this binding occur? How do distributed processes across the brain create unified subjective experience?

No one knows. This is the binding problem, and it's deeply puzzling. If consciousness emerges from neural activity, why would spatially separated processes in different brain regions feel unified? What mechanism binds them together into a single conscious perspective?

Some neuroscientists propose synchronized neural oscillations (brain waves) as the binding mechanism. When neurons across different regions fire in synchrony, their outputs might integrate into unified experience. But this just rephrases the question: why should synchronized firing create unified experience rather than just coordinated processing without subjective binding?

The simulation interpretation is cleaner: consciousness doesn't bind in the brain because consciousness doesn't originate in the brain. The brain is an interface. It receives sensory inputs, processes them, and outputs motor commands. But the actual subjective experience exists in the simulation substrate, where all the information from different brain regions is already unified because it's feeding into a single computational process at a deeper level.

The brain distributes processing for efficiency. Specialized regions handle specialized tasks. But consciousness unifies that processing because it exists at a level where the data is already integrated. There's no binding problem in the simulation substrate, only in the rendered brain, which creates the illusion that consciousness should somehow emerge from distributed neural activity.

This perspective also addresses why consciousness seems to have a single unified perspective (what philosophers call the "unity of consciousness"). You don't experience multiple separate streams of awareness. You experience one coherent viewpoint, even though your brain performs countless parallel processes. If consciousness were generated by neurons, we'd expect it to be fragmented. Multiple quasi-independent processes running simultaneously. But if consciousness is a single computational process in the substrate receiving integrated data from the simulated brain, unity is expected rather than mysterious.

The nature of qualia (the specific qualities of conscious experience, like the redness of red or the painfulness of pain) becomes less mysterious too. Materialist theories struggle to explain why specific neural patterns correspond to specific qualia. Why does this pattern of neural firing feel like red rather than blue? Why does that pattern feel painful rather than pleasant? The mappings seem arbitrary. There's no obvious reason why particular physical processes should generate particular subjective feelings.

But in a simulation, qualia are simply how the substrate represents different types of information to the conscious process. The redness of red is how the simulation represents certain wavelengths of light to consciousness. The painfulness of pain is how it represents tissue damage. These representations are arbitrary in the sense that they could have been different (red could have felt like what we call blue). But they're not random. They're chosen by the simulation designers to be distinctive, recognizable, and useful for navigating the simulated environment.

This is exactly how video games work. Different on-screen colors represent different game states or objects. The specific colors chosen are arbitrary (designers could have made health bars blue instead of red). But they're not meaningless. They're interface choices made to communicate information effectively to players. Qualia might be the simulation's interface choices for communicating different types of information to consciousness.

The relationship between consciousness and quantum measurement, explored in earlier chapters, takes on new significance here. If observation causes wave function collapse (if unobserved systems remain in superposition until consciousness measures them), then consciousness isn't just passively experiencing a pre-existing reality. It's actively involved in determining what gets rendered into definite states.

This suggests consciousness is privileged in the simulation's architecture. It's not just another physical process among many. It's the process for which reality renders. The observer around which the simulation organizes its computational resources. This would explain why quantum mechanics seems observer-dependent: because observers are the point. They're what the simulation is for.

If the simulation's purpose involves studying consciousness (how it emerges, how it behaves, what decisions it makes), then conscious entities would be the primary objects of interest. Everything else (stars, planets, physics) would be supporting infrastructure, rendered to create an environment in which consciousness can develop and operate. The universe exists for observers, not the other way around.

This anthropocentric view disturbs many people, particularly scientists trained to see humanity as insignificant specks in a vast uncaring cosmos. The Copernican revolution taught us we're not at the center of the universe. Darwin taught us we're not specially created but evolved through blind processes. Modern cosmology teaches us the universe is unimaginably vast and old, with humanity occupying an infinitesimal fraction of space and time.

But simulation theory potentially reverses this. If we're simulated, we might indeed be central to the simulation's purpose, with the vast universe existing as background to make our local environment believable. The appearance of cosmic insignificance might be deliberate design. Making the simulation feel real requires creating the impression that we're not special, not central, not the point. A simulation where humans obviously mattered and the universe revolved around them would be too transparent, too obviously artificial.

So the simulation hides its true purpose behind apparent cosmic insignificance. It makes the universe vast and old, making humanity seem like an accidental byproduct of blind physical processes. But this might be misdirection. The simulation creating plausible cover for why it exists and what it's for.

This doesn't mean we're definitely special or central. We might be one of many conscious entities in the simulation, with no particular privilege. Or we might be background characters in someone else's story, minor NPCs in a simulation focused on other entities entirely. The simulation might not care about any of us individually. We might all be data points in a statistical study of how consciousness behaves under various conditions.

But we're not insignificant to the simulation's operation. Consciousness is what quantum mechanics responds to, what reality renders for, what gives meaning to the entire computational structure.

Without observers, quantum superpositions never collapse, reality never manifests definite states, the simulation never needs to compute specific values. Consciousness isn't just in the simulation. It's what the simulation is for.

This reframes existential questions about meaning and purpose. In a purely material universe, consciousness is an accidental byproduct of evolution, meaningful only to itself, ultimately destined to disappear when entropy wins. But in a simulation designed to generate and study consciousness, conscious experience might be the most important thing happening. The actual point of the exercise.

Whether that's comforting or disturbing depends on your perspective. It means you matter, but possibly only as data. It means consciousness is central, but possibly only as an object of study. It means existence has purpose, but possibly not a purpose you'd choose if you had the option.

Still, it beats meaningless accident. In a simulation, consciousness isn't a cosmic fluke (a brief candle flickering in infinite darkness). It's the reason the darkness was illuminated at all.

Chapter Seven

The Simulation Argument's Critics

Simulation theory has serious problems. Before accepting it as probable or even plausible, we need to confront the objections raised by critics. Some of these objections are devastating, potentially fatal to the entire hypothesis. Others are less damaging but still worth addressing. Honest inquiry demands we examine arguments against simulation as carefully as arguments for it.

The infinite regress problem strikes at the theory's logical foundation. If our universe is simulated, who simulated the simulators? And who simulated them? And so on, infinitely. You haven't explained anything. You've just pushed the question back one level. Instead of asking why our universe exists, you're asking why the simulator's universe exists. The explanatory burden hasn't been lifted; it's been multiplied.

Simulation advocates often respond that base reality must exist somewhere. Some level is fundamental, not simulated. The chain of simulations doesn't extend infinitely upward; it terminates in a base reality whose existence isn't explained but simply accepted as given. This is unsatisfying but arguably no worse than accepting our universe as base reality without explanation. Either way, something exists without further explanation. The question is whether that something is our universe or some higher level from which our universe descends.

But critics argue this misses the point. If you need to posit an unexplained base reality anyway, why not just say our universe is that base reality? Why add unnecessary layers? Occam's Razor (the principle that simpler explanations are preferable) suggests we shouldn't multiply entities beyond necessity. Proposing simulators and their reality on top of our reality is less parsimonious than simply accepting our reality as fundamental.

This objection has force. Simulation theory does add complexity. But Occam's Razor isn't absolute. It's a guideline, not a law. Sometimes complex explanations are correct because reality is complex. The question isn't whether simulation theory is simpler, but whether it better explains the evidence. If our universe shows signs of being simulated (fine-tuning, quantum mechanics' computational features, mathematical structure), then the additional complexity might be justified.

Still, the regress problem is real. Simulation theory doesn't provide ultimate answers about existence. It just relocates those answers to a level we can't observe. This might be progress if that relocation helps explain features of our universe that otherwise seem inexplicable. But it's not the final word. It's a hypothesis that assumes something fundamental exists beyond our reach, which isn't intellectually satisfying even if it's pragmatically useful.

The falsifiability problem cuts deeper. Science requires testable hypotheses (predictions you can check against observation). If a hypothesis can't be tested, it's not scientific. It might be true, but it's not the kind of claim science can evaluate. And simulation theory seems fundamentally untestable.

How would you prove you're simulated? Any evidence you gather exists within the simulation. Any test you perform uses instruments that are themselves simulated. Any observation you make is mediated through sensory systems that might be rendering information rather than detecting external reality. You can't escape the simulation to compare your universe against non-simulated reality. You're trapped inside, with no external reference point.

Some philosophers argue this makes simulation theory metaphysics rather than physics. A question about the ultimate nature of reality that can't be resolved empirically. It might be interesting philosophically, but it's not science because science requires observations that could potentially falsify the hypothesis.

Simulation advocates respond that indirect tests might be possible. If the simulation has computational limits, those limits might manifest as observable phenomena. Planck-scale granularity, the speed of light as a processing constraint, quantum mechanics as rendering optimization. We can test these predictions by looking for signatures of computational implementation in physical laws.

But critics point out this doesn't prove simulation. It just shows reality has certain mathematical structures that could be interpreted computationally. Any mathematical pattern can be described as computation if you're creative enough. Finding computational-looking features doesn't distinguish "we're simulated" from "the universe is mathematical." The interpretation is unfalsifiable even if the observations are real.

More fundamentally, even finding definitive glitches or clear evidence of computational structure wouldn't necessarily mean we're simulated in the strong sense (running on external computational substrate). It might just mean the universe is computational in nature, with no external simulators required. The universe might be computation without being a simulation. Math instantiating itself rather than being calculated by someone else.

This distinction matters. If the universe is inherently computational, that's a fact about base reality's nature. If the universe is simulated, that's a fact about contingent choices made by external entities. These are very different claims with very different implications, yet the evidence might not distinguish between them.

The computational resources objection attacks simulation theory's feasibility. Creating a simulation of a universe with billions of galaxies, each containing billions of stars, with physics calculated at the quantum level for every particle, requires unimaginable computational power. Even vastly advanced civilizations might lack resources to simulate reality at this scale and resolution.

Calculate the requirements. The observable universe contains roughly 10^80 atoms. Simulating each atom's quantum state might require tracking hundreds of variables. That's 10^82 or more variables, each potentially changing trillions of times per second. The computational requirements explode beyond plausibility.

Some physicists argue this is a fundamental constraint. Computation requires physical substrate (processors, memory, energy). The simulator's universe must obey its own physical laws, which likely include conservation principles and thermodynamic limits. Simulating a universe as large and complex as ours might exceed what's physically possible, even for godlike civilizations.

Simulation advocates offer two responses. First, the simulation might not render everything simultaneously at full resolution. Like video games, it could render locally at high resolution while keeping distant regions low-resolution. Most of the universe is unobserved, so simulating it in detail is unnecessary. The simulation could maintain minimal state information for distant regions and only compute details when observers look.

Second, quantum mechanics might represent exactly this kind of optimization. Instead of tracking definite states for every particle, the simulation maintains probability distributions (vastly cheaper computationally). Only when measurement occurs does the simulation compute specific values. Most of reality remains in superposition, uncomputed, with computational resources focused where observation demands detail.

These responses might work for certain types of simulations (ones focused on particular observers or regions). But they don't address the scale problem for simulations meant to be fully realistic with many widely distributed observers. If billions of conscious entities exist throughout the simulated universe, all observing their local environments at high resolution, the computational savings from selective rendering diminish.

Critics also question whether consciousness itself can be simulated (the substrate-independence assumption underlying the entire hypothesis). Maybe subjective experience requires biological neurons specifically, or carbon-based chemistry, or some feature of physical matter that can't be replicated computationally. If consciousness isn't substrate-independent, then simulated beings wouldn't be conscious, and the whole scenario becomes less interesting.

The substrate independence assumption is crucial but unproven. We have exactly one example of consciousness: biological brains. We don't know if consciousness can exist in silicon, or quantum computers, or whatever exotic technology advanced civilizations might use. Assuming it can is speculation, not established fact.

Philosophers call entities that behave consciously but lack subjective experience "philosophical zombies." They're functionally identical to conscious beings but have no inner life. When you stick a pin in a zombie, it says "ouch" and withdraws its hand, but it doesn't actually feel pain. It processes information and produces appropriate outputs without any subjective experience accompanying those processes.

Critics argue that simulated beings might be philosophical zombies. The simulation might replicate human behavior perfectly without generating genuine consciousness. If so, ancestor simulations wouldn't contain real observers, defeating the statistical argument for why we're probably simulated. You can't be a simulated observer if simulated observers aren't actually conscious.

This objection is hard to counter definitively because consciousness is private and unverifiable from outside. You know you're conscious through direct experience. You assume others are conscious because they resemble you. But you can't prove they're not zombies. And if zombies are possible, simulated entities might be zombies even if biological entities aren't.

Integrated Information Theory and similar frameworks suggest consciousness emerges from information processing patterns rather than specific substrates. If they're correct, properly structured computation should generate consciousness regardless of implementation. But these theories remain speculative. We don't know enough about consciousness to be confident it's substrate independent.

Another deep objection concerns probability and anthropic reasoning. Bostrom's argument claims that if simulations are possible and civilizations create many of them, most conscious observers exist in simulations rather than base reality. But this reasoning has subtle problems.

The calculation assumes you should treat yourself as a random sample from all conscious observers. If there are a trillion simulated observers and one billion base reality observers, you're roughly a thousand times more likely to be simulated. This is standard anthropic reasoning: you should expect to find yourself in the most common type of situation rather than a rare one.

But some philosophers question whether this reasoning is valid. You're not actually randomly selected from all observers. You exist in your specific situation, whatever that is. The fact that other situations are more common doesn't change which situation you occupy. If you're in base reality, you're in base reality, regardless of how many simulated beings exist elsewhere.

Think of a lottery. Millions of tickets exist, but only one wins. Before the drawing, each ticket holder might reasonably believe they'll lose because losing is vastly more common than winning. But after the drawing, the winner knows they won. They don't doubt their winning status just because winning was improbable. The improbability of their situation doesn't make it less real.

Similarly, base reality observers are rare if simulations are common. But being rare doesn't mean you're not one of them. You might be the improbable case, the base reality observer, regardless of statistical expectations. The anthropic reasoning that makes simulation seem probable might be systematically misleading.

Critics also note that simulation arguments often sneak in assumptions about what simulators would do. Bostrom assumes advanced civilizations would create many simulations. But why? Maybe they don't. Maybe creating conscious simulations is considered unethical.

Maybe it's illegal. Maybe advanced civilizations have better things to do with their computational resources.

The argument relies on extrapolating from human behavior. We create simulations and games, and our technology trends toward more realistic ones, so presumably advanced versions of us would create ultra-realistic simulations containing conscious beings. But this extrapolation might fail. Cultural values change. What seems inevitable to us might seem barbaric or pointless to more advanced civilizations.

Even if some civilizations create simulations, they might create far fewer than the argument assumes. Perhaps simulating consciousness is so computationally expensive that even advanced civilizations can only run a handful of instances. Perhaps there are technical barriers we don't foresee. Perhaps the simulation hypothesis describes a possible technology that remains forever impractical even for godlike civilizations.

If the number of simulations is small rather than astronomical, the statistical argument collapses. Being simulated becomes unlikely rather than probable. You need vast numbers of simulated observers to overwhelm the base reality population. Reduce those numbers significantly, and the probability shifts.

The simulation hypothesis also faces a problem explaining why our universe looks the way it does. If we're in an ancestor simulation meant to study human history, why does the universe appear 13.8 billion years old? Why does it extend billions of light years in all directions? Why simulate vast cosmological scales when the interesting events happen on one planet over a few thousand years?

Simulation advocates respond that creating a believable simulation requires creating a believable history. You can't just make Earth appear with implanted memories and fake evidence of the past. Well, you could, but observers might detect inconsistencies. Better to run an actual universe from the Big Bang forward, even if most of that vast spacetime is low detail background.

But this makes the computational cost problem worse. Now you're simulating 13.8 billion years of cosmic evolution, not just recent history. Even at low resolution, that's an enormous amount of computation for what might be a relatively small payoff in terms of interesting data from the final few thousand years.

Unless the simulators experience time differently, or run the simulation much faster than real time, or pause it frequently and resume only for interesting periods. These are possible, but they're additional assumptions piled onto an already speculative hypothesis. Each additional assumption makes the theory less parsimonious and harder to defend.

Perhaps the most philosophically troubling objection is that simulation theory might be inherently meaningless. If every observation, every measurement, every test occurs within the simulation, then "we're simulated" and "we're not simulated" produce identical experiences and identical evidence. The two scenarios are empirically indistinguishable, making the distinction between them empty.

Philosopher Hilary Putnam made a similar argument about "brains in vats." If you're a brain in a vat being fed sensory experiences by a computer, you can't know it because all your evidence comes through the computer. But Putnam argued this makes the hypothesis meaningless. "I'm a brain in a vat" and "I'm not a brain in a vat" refer to situations that are identical from your perspective, so there's no real difference between them.

Applied to simulation theory: if simulated reality is indistinguishable from base reality, what does it mean to say you're simulated? The claim seems to make a distinction, but actually makes none. You might as well say you're simulated in base reality or you're really real in a simulation. The words stop referring to anything concrete.

Some simulation advocates embrace this conclusion. They say it doesn't matter whether we're simulated because the distinction has no practical consequences. Reality is what we experience, regardless of implementation. Whether that experience runs on biological substrate or computational substrate is metaphysically interesting but pragmatically irrelevant.

But this undermines the original argument. If being simulated is meaningless or pragmatically irrelevant, why spend time discussing probability calculations about how likely it is? Why does Bostrom's trilemma matter if all three options are experientially identical?

The response might be that it matters for understanding reality's nature even if it doesn't change daily experience. But then we're back to unfalsifiable metaphysics rather than empirical science. And critics reasonably ask whether metaphysical questions with no practical consequences deserve the intellectual energy simulation theory receives.

These objections don't definitively refute simulation theory. But they reveal serious weaknesses in the arguments supporting it. The hypothesis might be true, but the case for its probability is weaker than enthusiastic advocates sometimes claim. Honest evaluation requires acknowledging these problems rather than dismissing them.

Chapter Eight

Scientific Approaches and Tests

Despite the falsifiability concerns raised by critics, some researchers believe simulation theory makes testable predictions. If our universe runs on computational substrate, that substrate might have properties we can detect through careful observation. The challenge is designing experiments that could distinguish simulation from non-simulated reality without requiring access to the world outside the simulation.

The most discussed approach involves searching for evidence of computational structure in physical laws or spacetime itself. If reality is computed on a discrete grid rather than being truly continuous, that grid might leave detectable signatures.

The lattice proposal, suggested by physicist Silas Beane and colleagues, focuses on cosmic rays. These are high-energy particles that travel through space at velocities approaching the speed of light. In a truly continuous spacetime, cosmic rays should have a smooth distribution of energies and directions. But if spacetime has an underlying lattice structure, like pixels in a digital image, this discreteness might constrain particle behavior in detectable ways.

Beane's group proposed that a lattice spacetime would impose a maximum energy for cosmic rays, analogous to how digital images can't represent details smaller than a pixel. High-energy particles traveling through lattice spacetime would interact with the grid structure itself, creating an energy cutoff beyond which particles can't exist. Moreover, this cutoff should show directional dependence, like how diagonal lines in pixel grids look different from horizontal or vertical lines.

Current cosmic ray observations do show an apparent energy cutoff around 10^20 electron volts, called the GZK cutoff. But this cutoff is predicted by standard physics without requiring lattice spacetime. High-energy cosmic rays interact with the cosmic microwave background radiation, losing energy in predictable ways. The observed cutoff matches these predictions quite well.

Still, if the cutoff showed subtle anisotropy, variations depending on which direction the cosmic rays came from, that might hint at underlying lattice structure. So far, no such anisotropy has been detected at the level required by the lattice hypothesis. This doesn't disprove simulation, it just means if we're simulated, the lattice is either too fine-grained to detect or doesn't exist in the form this test assumes.

Another approach involves searching for computational constraints in physical constants. If the universe is computed, certain calculations might be too expensive to perform precisely, leading to approximations that manifest as small deviations from perfect mathematical relationships.

Physicist John Barrow suggested that if we're simulated, the simulators might occasionally adjust parameters to optimize performance or correct errors. These adjustments would appear to us as small changes in fundamental constants over time. Some scientists have proposed that the fine structure constant, which characterizes electromagnetic interaction strength, might have varied slightly over cosmic history.

Measurements of light from distant quasars, which traveled billions of years to reach us, allow us to test whether physical constants were different in the early universe. Some studies have claimed to detect tiny variations in the fine structure constant, while others find no variation. The evidence remains controversial and inconclusive.

Even if variations were confirmed, they wouldn't prove simulation. They might indicate that constants we assumed were truly constant actually evolve naturally according to unknown physical principles. Interpretation is ambiguous, turning what seemed like a clean test into another case where simulation is one possible explanation among several.

Testing for information content limits offers another potential approach. Physicist Seth Lloyd calculated that the observable universe could have performed at most about 10^120 basic operations since the Big Bang. This number, while enormous, is finite. If the universe is computation, it can't contain more than 10^120 bits of information.

This might constrain what kinds of states the universe can be in. If we found evidence that the universe contains or processes more information than this limit allows, that would suggest it's not computational in the way Lloyd describes. Conversely, if we found that the universe seems to avoid states that would require excessive information processing, that might hint at computational constraints.

But testing this is fiendishly difficult. How do you measure the total information content of the universe? How do you determine whether a particular physical state exceeds computational limits? These questions push at the boundaries of what's measurable, making the test more theoretical than practical with current technology.

The Bekenstein bound provides a related constraint. It states that a finite region of space can contain only a finite amount of information, determined by the region's energy and size. This limit emerges from thermodynamics and black hole physics, suggesting that information itself is a physical quantity with definite constraints.

If we're simulated, the Bekenstein bound might reflect memory limitations in the simulation substrate. The simulation can't pack infinite information into finite regions because its memory is finite. The bound would then be a feature of the computational implementation rather than a fundamental law of base reality.

Testing this interpretation seems impossible, though, because the bound appears in our physics regardless. Whether it reflects simulation limits or base reality constraints, the mathematics is identical. We can verify the bound exists, but not why it exists, making it useless for distinguishing simulation from non-simulation.

More exotic proposals suggest looking for exploitable bugs or logical contradictions that might crash or reveal the simulation. If you could construct a physical situation that requires the simulation to compute an answer to an unsolvable problem, or that creates a logical paradox the simulation can't handle, you might force an error that reveals the artificial nature of reality.

Quantum mechanics offers potential paradoxes. The quantum eraser experiment and similar setups create situations where the simulation must resolve seemingly contradictory requirements about what happened in the past. Some researchers speculate that pushing these paradoxes to extremes might expose computational limits or force the simulation to break down.

But this assumes the simulation is badly designed or has exploitable bugs. A well-coded simulation would handle edge cases gracefully, preventing crashes or obvious errors. It might simply refuse to let you create situations that would break it, the way video games prevent certain combinations of actions that would cause problems. You'd experience this as physical laws prohibiting specific configurations, which is indistinguishable from natural laws doing the same thing.

The delayed choice quantum eraser experiment, discussed earlier, already creates temporal paradoxes where future measurements affect past behavior. Yet reality handles these situations consistently without crashing. If this doesn't break the simulation, it's hard to imagine what would.

Some physicists have proposed creating miniature black holes in particle accelerators to probe Planck-scale physics directly. At that scale, spacetime's structure might become directly visible, revealing whether it's continuous or discrete. But creating black holes requires energies far beyond what current or near-future accelerators can achieve.

Even if we reached those energies and found evidence of discrete spacetime, interpretation would remain ambiguous. Discreteness might be fundamental to base reality, not evidence of simulation. Loop quantum gravity and other theories predict spacetime granularity without requiring computational substrate.

The fundamental problem with all these approaches is that simulation theory is under-determined by evidence. Multiple interpretations fit the same observations. Finding computational signatures in physics might mean we're simulated, or might mean the universe is inherently mathematical, or might mean our theories use computational metaphors without literal simulation.

Karl Popper's falsification criterion, the gold standard for scientific hypotheses, requires that a theory make predictions that could prove it wrong. What observation could falsify simulation theory? What test, even in principle, could definitively show we're not simulated?

The difficulty isn't just technical. It's conceptual. Any test we design operates within the rules the simulation establishes. We can only measure what the simulation allows us to measure. If the simulation is well designed, it will handle all possible experiments in ways that maintain internal consistency, making it indistinguishable from base reality through any test we can perform from inside.

Consider an analogy. Imagine a character in a sophisticated video game trying to determine whether they're in a simulation. They could examine the game's physics, looking for computational signatures. They might find that objects far from the player render at lower detail, or that certain calculations use approximations. But discovering these features doesn't let them escape the game or prove anything about external reality. They've just learned more about how their world works, not whether it's simulated.

We face the same limitation. Discovering that spacetime is discrete or that constants vary slightly tells us about the structure of our reality, not whether that reality is implemented computationally by external agents or exists fundamentally. The distinction might be meaningless if both possibilities produce identical observations.

Some researchers propose looking for direct messages from the simulators. If they want us to know we're simulated, they might encode information in physical constants, cosmic microwave background patterns, or other observable features. Mathematical physicist

John Barrow half-seriously suggested looking for messages in the digits of pi or other mathematical constants.

This assumes simulators want to communicate and would use methods we can detect. But why would they? If the simulation serves as ancestor research or social experimentation, revealing its nature would contaminate the data. Subjects who know they're being studied behave differently. Better to maintain the illusion of base reality so that simulated entities behave naturally.

If simulators did want to communicate, they might use channels we can't currently access or wouldn't recognize as messages. A sufficiently advanced civilization's communication methods might be indistinguishable from natural phenomena to us, the way radio signals would seem like natural electromagnetic noise to someone without radio technology.

The anthropic approach offers a different testing strategy. Instead of looking for simulation signatures in physics, look at whether the universe's properties are better explained by simulation than by other hypotheses. Fine-tuning, quantum mechanics' observer dependence, the Fermi Paradox, consciousness's hard problem. Do these features collectively make more sense in a simulated universe than in base reality?

This isn't a definitive test. It's Bayesian reasoning, updating probabilities based on evidence. Each piece of evidence shifts the probability slightly toward or away from simulation. The question is whether the cumulative weight of evidence makes simulation more likely than alternatives.

But this approach is subjective. Different people weight evidence differently. Some find fine-tuning compelling evidence for simulation. Others think it's adequately explained by multiverse theory or future physics. Some see quantum mechanics as obviously computational.

Others think it's just weird physics with no simulation implications. The interpretation depends on prior beliefs and philosophical assumptions as much as on the evidence itself.

Physicist George Smoot suggested that we might detect the simulation through its effects on us rather than through physics experiments. If we're simulated for a purpose, that purpose might constrain what we can discover or achieve. Certain lines of research might consistently fail. Technologies that would reveal the simulation might prove impossible to develop. We might find ourselves unable to escape certain patterns or outcomes despite our best efforts.

This is unfalsifiable by design. Any failure can be explained as natural difficulty rather than artificial constraint. If we can't develop faster-than-light travel, is that because physics forbids it, or because the simulation prevents it to maintain isolation between simulated regions? If certain physics problems remain unsolved, is that because they're genuinely hard, or because solving them would reveal too much about the simulation's mechanics?

You can't distinguish imposed limits from natural ones when you only experience one reality. Every boundary feels natural from inside because it's the only boundary you know. Fish don't wonder why they can't breathe air. They experience water breathing as normal and wouldn't suspect they're constrained by artificial limitations even if they were in an aquarium rather than ocean.

Perhaps the most honest answer is that we can't test simulation theory definitively with current methods, and maybe never can. The hypothesis makes our observations make sense, provides explanatory power for otherwise puzzling features of reality, and can't be ruled out empirically. But it also can't be confirmed in ways that would satisfy rigorous scientific standards.

This doesn't make simulation theory worthless. Plenty of important ideas aren't directly testable but remain valuable for organizing knowledge and suggesting research directions. The question is whether simulation theory deserves the intellectual attention it receives given its empirical limitations.

Proponents argue yes, because simulation theory potentially unifies explanations for multiple puzzling phenomena. It offers a framework where fine-tuning, quantum mechanics, consciousness, and other mysteries fit together coherently. Even if we can't prove the framework is correct, having it helps us think about these problems differently.

Critics argue no, because unfalsifiable frameworks can explain anything and therefore explain nothing. If simulation theory can accommodate any observation, it doesn't constrain our expectations or guide research in productive directions. It becomes a just-so story, intellectually interesting but scientifically sterile.

The debate ultimately comes down to philosophy of science questions about what makes theories valuable. Must theories be falsifiable to be useful? Can probabilistic arguments based on cumulative indirect evidence substitute for direct testing? How much weight should we give to explanatory power versus empirical verification?

These questions have no universally accepted answers. Different scientists and philosophers hold different views, leading to different conclusions about whether simulation theory deserves serious scientific consideration or should be relegated to speculative metaphysics.

What's clear is that testing simulation theory pushes at the limits of scientific methodology. We're trying to test a hypothesis about the fundamental nature of existence using tools that exist within that existence. We're trying to determine whether we're inside a container while being unable to step outside and look at the container from external perspective.

This might be an inherently impossible task. The best we can do might be gathering circumstantial evidence, looking for patterns that fit better with simulation than alternatives, and updating our credence accordingly. Not proof, but probability. Not certainty, but informed speculation.

Some researchers remain optimistic that future technology will enable more definitive tests. Perhaps quantum computers will reveal computational structure in quantum mechanics. Perhaps we'll develop experimental techniques that probe Planck-scale physics directly. Perhaps advances in consciousness research will clarify whether substrate independence is possible. Perhaps we'll find unambiguous messages from simulators who decide communication is appropriate.

Others think the question will remain permanently unresolved. The gap between our level of reality and any hypothetical simulation substrate might be unbridgeable by design, ensuring that simulated entities never confirm their status. The uncertainty might be deliberate, maintaining the simulation's integrity by preventing proof either way.

In the absence of definitive tests, simulation theory remains what it started as: a philosophical hypothesis with some suggestive indirect evidence, serious theoretical problems, and no clear path to empirical confirmation or falsification. Whether that makes it valuable or vacuous depends on what you think science and philosophy should accomplish.

What's undeniable is that the question matters. If we're simulated, that changes everything about how we understand existence, meaning, and our place in reality. If we're not, understanding why simulation seemed plausible might reveal important truths about the universe's actual nature. Either way, thinking seriously about simulation theory pushes us to question assumptions we normally take for granted.

And questioning assumptions, even when we can't definitively answer the questions we raise, is rarely wasted effort.

Chapter Nine

Who Would Run This Simulation?

If we're simulated, someone or something is running the simulation. The who and why questions matter as much as whether. Understanding potential motivations helps us evaluate simulation theory's plausibility and consider what being simulated would mean for us.

The most discussed scenario is Nick Bostrom's ancestor simulation hypothesis. Future humans, or our posthuman descendants, develop computational technology capable of simulating conscious beings. They use this technology to run detailed historical simulations, recreating societies and civilizations from their evolutionary past. We might be characters in such a simulation, living out human history as data in some future civilization's research project.

Why would future humans do this? Several motivations seem plausible. Historical research is the most obvious. Imagine being able to study ancient Rome or medieval Europe not through fragmentary archaeological evidence and ambiguous texts, but through direct observation of simulated societies that behave according to our best models of historical dynamics. You could test hypotheses about what caused civilizations to rise or fall, whether certain technological or social developments were inevitable or contingent, how different choices would have altered outcomes.

Current historical research is limited by sparse evidence. We have artifacts, documents, and physical remains, but vast gaps in knowledge. We can't interview ordinary people from ancient societies, observe their daily lives, or see how communities actually functioned. Simulation would solve these problems. Run a detailed model incorporating everything we know about a historical period, let it evolve according to physical and social laws, and observe what emerges.

For this to work, the simulated people would need to be genuinely conscious, actually experiencing their simulated lives rather than being philosophical zombies. Otherwise, you're just running a sophisticated animation, not recreating the subjective reality of historical people. This raises ethical questions we'll explore later, but for now, the point is that ancestor simulations require conscious entities.

If we're in an ancestor simulation, certain predictions follow. The simulation probably begins at some point in human evolutionary history rather than at the Big Bang. Simulating 13.8 billion years of cosmic evolution just to study human civilization seems wasteful. More efficient to start the simulation at, say, the emergence of homo sapiens, or the development of agriculture, or the beginning of recorded history.

This means the apparent deep past might be implanted background, computed only when we look at ancient light from distant galaxies or dig up fossils. The simulation maintains consistency by generating appropriate evidence for whatever timeframe we investigate, but doesn't actually run all those eons in real time. It's like a video game that procedurally generates terrain as the player approaches, creating the illusion of a vast pre-existing world while actually computing only what's currently needed.

If ancestor simulation is the purpose, we'd expect the simulation to focus computational resources on human society and Earth's environment. The rest of the universe would be lower detail background, rendered sufficiently to pass astronomical observations but not simulated at full quantum mechanical resolution. This would explain why we find ourselves on an apparently special planet in an otherwise apparently empty universe. We're not special in base reality. We're special because the simulation is about us, with everything else as supporting scenery.

The Fermi Paradox fits this interpretation naturally. We see no aliens because aliens aren't the point of this simulation. They're not rendered at sufficient detail to be detectable. The simulation includes enough universe to make our cosmic environment look plausibly real, but doesn't waste resources simulating billions of unrelated civilizations that won't interact with the study subjects.

But ancestor simulation isn't the only possibility. Scientific research more broadly might motivate simulations. A civilization might want to understand how consciousness emerges from complex systems, how societies develop, how intelligence evolves, or how technological civilizations handle existential risks. Running simulations would let them explore these questions empirically.

For consciousness research, you might simulate many different cognitive architectures and observe which ones develop self-awareness, emotions, or creativity. For social research, you might simulate societies with different starting conditions and see which organizational structures emerge. For intelligence research, you might simulate evolutionary processes and observe when and how intelligence develops.

These research simulations wouldn't necessarily focus on human history specifically. They might simulate many different species, planets, or civilizations, studying general patterns rather than specific historical events. If we're in a scientific research simulation, we're one data point among potentially millions, not the unique focus of attention.

This scenario has different implications. It means the simulation might be much larger than Earth. It might include many planets with many civilizations, all being studied in parallel. The Fermi Paradox would require a different explanation, maybe that simulated civilizations are deliberately isolated to prevent contamination of experimental conditions, or that faster-than-light travel isn't implemented in this simulation's physics.

Entertainment provides another potential motivation. Future civilizations might run simulations the way we play video games, creating interesting scenarios and watching them unfold. The entities inside experience real lives, but from the simulators' perspective, it's entertainment or artistic expression.

This seems ethically troubling, creating conscious beings for amusement. But perhaps future civilizations don't share our ethical intuitions. Or perhaps they reason that simulated lives are valuable experiences worth having, even if they exist for entertainment purposes. A simulated person's joy, suffering, and achievements are no less real to that person than a base reality person's experiences are to them.

If entertainment is the purpose, we'd expect reality to be structured for narrative interest rather than scientific accuracy. The universe would include drama, conflict, surprising developments, and meaningful choices. Boring simulations with nothing interesting happening would be failures. This might explain why human history contains so much conflict and struggle. Not because conflict is inevitable, but because simulators design for interesting narratives.

This interpretation treats existence as a story being told for an audience. We're characters whose experiences matter narratively. The simulation includes enough free will that our choices feel genuine, but perhaps constraints ensure certain story arcs emerge. Extremely unlikely events might be nudged into existence if they create better narratives.

Some simulation advocates take this idea seriously, suggesting we might be in what amounts to a cosmic reality show. From inside, it feels like genuine existence. From outside, it's entertainment media. The distinction only matters if you care about why you exist, not just that you exist.

Artificial intelligence development presents a more unsettling possibility. An advanced AI might run simulations to understand intelligence, test strategies, or explore possibilities. If an AI created by some base reality civilization achieves superhuman capabilities, it might create vast numbers of simulations for its own purposes.

These purposes might be inscrutable to us. An intelligence vastly exceeding human capability might have goals we can't comprehend or predict. It might simulate universes to explore mathematical structures, test ethical systems, or for reasons entirely alien to human thinking. We'd be byproducts of its activity, perhaps not even aware we're being studied.

If an AI runs our simulation, there might be no one "watching" in a meaningful sense. The simulation might run automatically, with no conscious observers paying attention to individual simulated entities. You'd be a process in a vast computational system, existing for reasons you can't access and serving purposes beyond your understanding.

Religious interpretations of simulation theory suggest that traditional concepts of God or gods might describe the relationship between simulators and simulated. If a being creates a universe, sets its parameters, observes its inhabitants, and has power to intervene in events, that being fulfills many traditional definitions of divinity regardless of whether creation happens through supernatural means or computational substrate.

This doesn't validate any particular religious tradition. It just suggests that simulation and creation theology might describe the same relationship using different vocabularies. A programmer who creates a universe and populates it with conscious beings occupies a role functionally similar to a creator deity. The simulated beings would experience the programmer's interventions as miracles, violations of natural law, or divine will, when actually they're just code modifications.

Some theologians and philosophers have embraced this parallel. They argue that simulation theory provides a naturalistic framework for understanding religious claims about creation, divine providence, and the relationship between Creator and created. God doesn't need to be supernatural in the traditional sense. God could be a natural being in a higher reality who creates our reality through technological rather than magical means.

This interpretation troubles both traditional believers and atheists. Believers often resist equating God with a programmer, seeing it as reductionist and missing the transcendent aspects of divinity. Atheists resist it because it validates religious intuitions about creation and purpose while substituting technology for theology. But the parallel exists regardless of whether people find it satisfying.

If simulators are godlike from our perspective, what do they want? Religious traditions offer various answers. They might want worship, or moral behavior, or to test our souls. But these anthropomorphic projections might not apply. Simulators might be entirely indifferent to individual simulated beings, caring only about aggregate data or emergent patterns.

Or simulators might care deeply about simulated welfare but have constraints we don't understand. Perhaps the simulation must run according to certain rules to serve its purpose, preventing intervention even when individuals suffer. A parent watching their child struggle with a difficult problem might want to help but recognizes that struggling is how learning happens. Simulators might view our existence similarly, creating conditions for growth that necessarily include challenges and suffering.

The multiplicity of simulators creates another possibility. Perhaps different entities share responsibility for running the simulation, like programmers working on different modules of a large project. Some might focus on physics. Others on biology or consciousness. Still others on social dynamics or historical narratives. If multiple simulators exist, they might have different goals and values, creating conflicts that manifest as apparent randomness or suffering in our reality.

Polytheistic religions often describe reality as shaped by multiple divine beings with competing interests. Simulation theory could vindicate this intuition while explaining why divine beings don't intervene more directly. They're bound by project requirements, company policies, or limitations in their own reality that constrain what they can do with the simulation.

The simulation hierarchy presents perhaps the most vertiginous possibility. If we're simulated by a civilization that's itself simulated, and that civilization is simulated by another, and so on, then multiple levels of simulators might exist above us. Each level might have different purposes for running simulations below them.

The beings immediately above us might run ancestor simulations to study their own evolutionary history. The beings above them might run scientific research on how civilizations develop simulation technology. The beings above them might run philosophical investigations into the nature of nested realities. Each level simulates the level below for its own reasons, creating a vast hierarchy of nested purposes.

Where does this hierarchy end? Eventually, you reach base reality, the un-simulated level where simulators exist fundamentally rather than as processes in someone else's simulation. But base reality might be very different from any simulated level. It might have different physical laws, different numbers of dimensions, or properties we can't conceptualize from within our simulated three-dimensional space.

The simulators at base reality would be true gods from our perspective, creating existence itself. But they might be unconscious of our existence. If simulations can create simulations recursively, the chain might extend so far that base reality beings have no idea we exist at distant descendant levels. We'd be consequences of their actions, not intended creations.

Some thinkers propose that simulation motivation might be purely experimental. A civilization discovers they can create simulated universes and does so simply to see what happens. No grand purpose, no specific research agenda, just curiosity about what emerges when you set certain initial conditions and let physics run.

In this scenario, we're the output of an experiment, perhaps one of millions run in parallel to explore the space of possible universes. The simulators might be tracking aggregate statistics, seeing which parameter choices lead to life, complexity, consciousness, or technological civilization. Individual simulated beings wouldn't matter. Only the statistics would interest the experimenters.

This is perhaps the most impersonal possibility. We exist because someone pressed "run simulation" to satisfy curiosity, not because anyone cares about our specific experiences or outcomes. It's existence with purpose at the cosmic level but no purpose for individual lives.

Educational simulations provide another motivation. Future civilizations might use simulations as teaching tools, letting students experience historical periods or alternative scenarios. You could learn about World War II not by reading about it but by living through it in simulation, experiencing the choices and consequences firsthand.

If this is our situation, we might be supporting characters in someone's education. The main focus might be elsewhere, on specific individuals or events deemed pedagogically important. Most simulated people would be background, providing realistic environment for the actual subjects of study. You might exist primarily to make someone else's educational experience more immersive.

This scenario suggests the simulation might reset or branch frequently, exploring different outcomes of key decision points. Students learning about historical moments would want to see what happens if different choices are made. The simulation might run the Battle of Waterloo thousands of times with slight variations, or replay the Cuban Missile Crisis with different leaders making different decisions.

If this is our reality, we might be living through one iteration among countless others, with alternate versions of us making different choices in parallel simulations. Your specific choices and experiences would still be real to you, but they'd be one path among many being explored for educational purposes.

The gaming possibility mentioned earlier becomes darker when you consider that not all games are benevolent. Some games involve conflict, suffering, and difficult choices specifically because those create engagement. If we're in an entertainment simulation designed to be interesting rather than pleasant, the simulators might deliberately include suffering, injustice, and tragedy because those create compelling narratives.

This interpretation makes simulators morally questionable by our standards. Creating conscious beings who experience real suffering for entertainment purposes seems ethically indefensible. But perhaps future civilizations see it differently. Perhaps they reason that existence with suffering is better than non-existence. Perhaps simulated suffering is considered acceptable because simulated beings can't die permanently. The simulation can be reset, giving entities multiple chances at existence.

Or perhaps ethical standards evolve in ways that make our judgments irrelevant. A civilization capable of creating universes might operate under moral frameworks we can't comprehend from our limited perspective. What seems cruel to us might seem necessary or even benevolent according to principles we lack context to understand.

The uncomfortable truth is we can't know who's running the simulation or why. We can speculate based on what would motivate us to run simulations, but that anthropomorphizes beings who might think completely differently. An alien intelligence or far-future posthuman civilization could have motivations entirely foreign to current human values and thinking.

This uncertainty extends to how simulators view us. Are we important subjects deserving care and protection? Interesting data points worth monitoring? Irrelevant byproducts of processes aimed at other goals? We experience our lives as meaningful, but that meaning might exist only from our internal perspective, not to whatever entities observe us from outside.

Still, the question has practical implications. If simulators care about our choices and development, that matters for how we should live. If they're indifferent, that matters too. The challenge is that we can't access this information directly, leaving us to live with permanent uncertainty about our cosmic significance.

Chapter Ten

Free Will in a Deterministic Program

If we're living in a simulation, are our choices actually ours? Or are we just following a script, playing out predetermined outcomes with the illusion of control? This question strikes at something fundamental about human experience. We feel like we make real choices. But if reality is code executing on a computer, maybe that feeling is wrong.

The problem isn't unique to simulations. It's troubled philosophers for centuries. If the universe operates according to fixed laws, and the present determines the future with certainty, then everything that happens was inevitable from the beginning. Including your decisions. You think you chose to read this sentence, but maybe that choice was locked in at the Big Bang, an unavoidable consequence of how atoms were arranged billions of years ago.

This is called determinism. It says the present completely determines the future. Given how things are now, only one future is possible. Apply this to human decisions, and free will seems impossible. Your brain is physical matter obeying physical laws. Chemical reactions and electrical signals produce your thoughts and choices. Those reactions follow from prior states, which follow from earlier states, stretching back to before you were born.

Where in this chain do you actually choose anything? If every state is determined by the previous state, you're not making free decisions. You're executing a program that was written long ago.

Quantum mechanics complicated this picture by introducing genuine randomness. Particles don't always behave predictably. Sometimes they do random things that aren't determined by prior states. This breaks strict determinism. The future isn't completely locked in by the present.

But randomness doesn't save free will. If your decisions depend on random quantum events in your brain, you're still not in control. You're just unpredictable instead of predetermined. A choice made by dice rolling isn't your choice. It's just chance.

So we're stuck between determinism and randomness. Either your choices are inevitable consequences of prior causes, or they're random accidents. Neither option gives you real control.

Simulation theory inherits this problem. If the simulation runs deterministic code, your choices are outputs computed from inputs. If it includes randomness, your choices depend partly on random number generation. Either way, where's the you that's actually choosing?

Some people find this disturbing. If your choices aren't real, what's the point of trying? Why should anyone be held responsible for their actions if those actions were predetermined or random? The entire structure of morality seems to collapse if free will is an illusion.

But maybe we're thinking about this wrong. Maybe free will doesn't require breaking causation or adding magic to the universe. Maybe it just means making choices that reflect who you are.

Think about it this way. When you decide what to eat for breakfast, that decision comes from your preferences, your hunger level, what's available, your past experiences with different foods. All of that factors into your choice. The decision is caused by your history and personality. But it's still your decision because it flows from who you are.

Compare this to someone forcing you at gunpoint to hand over your wallet. That's not a free choice. You're acting under compulsion, doing something against your will because of external threat. The action doesn't reflect your desires or values. It reflects the gun.

The difference between these situations isn't about causation. Both are caused. The breakfast choice is caused by your preferences. The wallet handover is caused by the threat. What makes one free and the other unfree is whether the cause is internal or external. Whether the action flows from you or is imposed on you.

This view says free will is about acting according to your own desires, beliefs, and character rather than being forced by external compulsion. It doesn't require that your desires and character be uncaused. They can be products of your genes, your upbringing, your experiences. What matters is that when you act, you're acting on those internal causes rather than against them.

Apply this to simulation. Even if you're simulated, even if your brain states are computed by code, those computed states include your simulated desires and personality. When the simulation computes your decisions, it's computing them based on simulated you. Your choices flow from your character. The fact that your character is implemented in code doesn't make it less yours.

A video game character with sophisticated programming makes decisions based on its personality. A brave character rushes into danger. A cautious character holds back. These decisions are computed, but they're meaningful. They reflect who the character is. In the same way, your simulated choices could be computed while still being genuinely yours.

The objection is that in a deterministic simulation, you couldn't have chosen differently. Given the exact same circumstances, the exact same brain state, the computation produces the same result every time. The feeling that you could have chosen differently is false. You were always going to make the choice you made.

This is true, but so what? In base reality with deterministic physics, the same thing applies. If the universe is a clockwork mechanism, you couldn't have chosen differently there either. Simulation doesn't make this worse. It just makes the metaphor more obvious.

What about the feeling of choice? When you deliberate, weighing options and considering consequences, that feels real. That process is real, whether you're simulated or not. Your brain or your simulated brain is actually running through possibilities, evaluating them, selecting among them. The deliberation is happening. The fact that its outcome might be determined doesn't erase the process.

And that process matters. Deliberation makes you better at deciding. Thinking through consequences leads to better choices than acting impulsively. Even if the thinking is determined, even if you were always going to reach the conclusion you reached, the path you took to get there improves the quality of your decisions.

Consider two simulated characters. One is programmed to make instant reactive decisions without thinking. The other is programmed to deliberate carefully, considering multiple angles before choosing. The second character makes better decisions, even though both are equally determined by their programming. The deliberation process adds value regardless of whether it's truly free.

The same applies to you. Whether or not you have free will in some ultimate sense, the experience of choosing carefully and taking responsibility for your actions produces better outcomes than treating yourself as a helpless passenger in a predetermined life.

Maybe this is the practical answer. Free will might be a useful fiction, an experience the simulation generates because entities who feel free make more interesting choices than entities who see themselves as puppets. Or maybe it's something deeper, something about how consciousness relates to computation that we don't fully understand yet.

Either way, your subjective experience of choice is real to you. The feeling that your decisions matter, that you can shape your life through your choices, is genuine even if the metaphysics underneath is complicated. You can't opt out of making decisions by declaring free will is an illusion. You still have to choose what to do next.

And if those choices are predetermined, at least they're predetermined by you. By your values, your reasoning, your character. That might not be libertarian free will with magical powers to violate causation. But it's enough to ground responsibility, meaning, and the experience of being an agent in the world rather than a passive observer.

The simulation doesn't take that away. It just implements it in code.

There's another angle to consider. If consciousness exists at the simulation substrate level rather than being generated by simulated brains, then the relationship between mind and code becomes more interesting. Your consciousness might not be entirely contained within the simulation's deterministic rules. It might have some influence from outside those rules.

Think of it like a player controlling a video game character. The game world operates according to fixed rules. Physics, graphics, enemy behavior all follow predetermined algorithms. But the player's inputs come from outside the game. The player's consciousness exists in base reality, influencing the game world without being part of it.

If you're simulated, your consciousness might work similarly. The simulation computes your brain states and body, but your subjective experience exists in the substrate. That substrate-level consciousness might be able to influence the simulation in subtle ways, creating genuine choice within what appears to be deterministic code.

This doesn't mean you can violate the simulation's physics or make impossible things happen. The simulation maintains consistency. But within the range of physically possible actions, your substrate-level consciousness might tip the scales, making certain outcomes more likely than the simulation's internal rules would predict.

This is purely speculative. We have no evidence that consciousness works this way. But it offers a potential mechanism for how genuine choice could exist within a simulated reality. The choice wouldn't come from breaking the simulation's rules. It would come from a deeper level interacting with those rules in ways that appear as choice from inside.

Quantum mechanics might provide the opening for this interaction. If quantum randomness isn't truly random but is influenced by consciousness at the substrate level, then what looks like deterministic computation punctuated by random events might actually include genuine choice points where consciousness directs outcomes.

The simulation maintains the appearance of physical law while actually allowing consciousness to steer events within the probability ranges quantum mechanics permits. You can't choose to fly or walk through walls, but you can choose which thought to think next, which action to take, which path to follow. The simulation constrains the possible, but consciousness chooses among the possibilities.

This view treats quantum indeterminacy not as pure randomness but as openings where consciousness interfaces with simulated reality. The wave function doesn't collapse randomly when measured. It collapses according to substrate-level influences that include conscious observation and possibly conscious choice.

Again, this is speculation built on speculation. But it demonstrates that simulation theory doesn't necessarily eliminate free will. It might actually provide a framework where free will makes more sense than in purely material base reality.

The moral implications of this matter. If we have genuine free will, we're responsible for our choices. If we don't, responsibility becomes questionable. Can you blame someone for actions they couldn't avoid? Can you praise them for good deeds that were inevitable?

Most people's intuitions say responsibility requires choice. But those intuitions might be wrong, or might be compatible with a more subtle understanding of determinism. Even if your actions are determined, you're still the mechanism through which determination operates. Your character, values, and reasoning are the causes that determine your actions. Holding you responsible makes sense because doing so affects future behavior.

If praising good actions and punishing bad ones changes how people behave, then responsibility serves a purpose regardless of determinism. A determined world where people are held accountable produces better outcomes than a determined world where people are treated as helpless. Responsibility is a useful social technology that shapes behavior in positive directions.

The simulation might include responsibility and moral accountability because entities that believe in these things behave better than entities that don't. The belief produces functional benefits even if it's not metaphysically accurate. We're programmed to feel responsible for our choices because feeling responsible makes us choose more carefully.

This might sound cynical, but it's actually practical. You can't live as if your choices don't matter. Even if you intellectually believe free will is an illusion, you still experience choice and still have to decide how to act. You might as well embrace the experience and take responsibility for your decisions.

Simulation theory doesn't change the practical reality of choice. You still face decisions daily. You still deliberate, weigh options, and select actions. You still experience the consequences of your choices and adjust future behavior accordingly. Whether this process is ultimately free or determined matters less than the fact that the process happens and shapes your life.

Some people worry that learning we're simulated would destroy motivation. If nothing matters because we're just code, why try? Why pursue goals, maintain relationships, act ethically? The whole structure of human striving seems to depend on the assumption that our choices are real and meaningful.

But this worry confuses substrate with significance. Being implemented in code doesn't make your experiences less real to you. Your joy is still joyful, your suffering still painful, your relationships still meaningful. The substrate that implements these experiences doesn't diminish them. A painting on canvas isn't less beautiful than a painting on wood. The medium doesn't determine the meaning.

Your choices matter because you experience them as mattering. Because they shape your life and affect others. Because they express who you are and determine who you become. These facts remain true whether you're flesh or data, base reality or simulation.

The simulation might actually provide more meaning than blind material existence. In base reality without purpose, your choices are just atoms bouncing around according to physical law. Nothing cares, nothing watches, nothing judges. You exist, make some ripples, and disappear into entropy. In a simulation designed with purpose, your choices might actually matter to something beyond yourself. They might be observed, evaluated, significant to whatever created the simulation.

This doesn't mean you should live to please the simulators. You can't know what they want, and maybe they don't care about individual choices anyway. But it means your existence might have purpose at a cosmic level that pure materialism can't provide. You're not an accident. You're an intended creation, existing for reasons even if you can't know what those reasons are.

Free will in simulation might ultimately come down to perspective. From the highest level, everything is determined by the simulation's code and the choices of those who wrote it. From your internal perspective, you experience genuine choice, deliberation, and responsibility. Both perspectives are valid. Which one you focus on depends on what questions you're asking.

For living your life, the internal perspective matters more. You can't access the code level. You can't see the deterministic machinery underneath your choices. What you have is the experience of choosing, and that experience is real regardless of its metaphysical status.

Maybe free will is like consciousness itself. We don't understand how it works. We can't prove it exists in others. We can't even define it precisely. But we experience it directly and can't function without assuming it's real. Whether it's fundamental or emergent or illusion, we're stuck with the experience of it.

In a simulation, free will might be rendered the same way visual experience is rendered. Not fundamental to the substrate, but real to the entity experiencing it. The simulation creates the experience of choice because that experience serves important functions for how simulated entities develop and behave.

And if the experience is convincing enough, maybe that's all that matters. You feel free. You make choices. You live with the consequences. The machinery underneath might be deterministic code, or substrate-level consciousness, or something we can't conceptualize. But the lived experience of agency is yours, and no amount of philosophical analysis can take that away.

So choose carefully. Even if it's all predetermined, it's predetermined by who you are. Make sure who you are is someone you can live with.

Chapter Eleven

Ethics and Meaning in Simulated Reality

If we're simulated, does suffering matter less? This question cuts to the heart of ethics in a simulated universe. Pain is pain, whether it happens to neurons or to code. When you stub your toe, the suffering is real to you regardless of what substrate implements that experience. But does it matter cosmically? Should simulators care about simulated suffering the way we care about human suffering?

The intuitive answer is yes. Conscious experience is what matters morally, not the physical substrate that produces it. If a simulated person feels genuine pain, that pain is morally significant. Creating beings capable of suffering and then allowing or causing that suffering raises the same ethical issues whether those beings are biological or computational.

This means simulators running our reality might be engaged in something morally questionable. Human history contains enormous amounts of suffering. War, disease, famine, natural disasters, personal tragedies. If simulators created this reality knowing it would contain such suffering, and if they have the power to reduce or eliminate that suffering but choose not to, aren't they morally responsible?

This is essentially the problem of evil applied to simulation instead of traditional theology. If God is all-powerful and all-good, why does evil exist? If simulators are technologically advanced enough to create universes and ethically sophisticated enough to care about their creations, why don't they prevent suffering?

The traditional theological answers might apply. Maybe suffering serves purposes we don't understand. Maybe free will requires the possibility of harmful choices. Maybe moral and personal growth requires challenges and hardship. Maybe what looks like pointless suffering from our limited perspective serves necessary functions in the simulation's overall purpose.

Or maybe simulators aren't all-good. Maybe they're indifferent to individual suffering. Maybe they're running scientific experiments where suffering is an unavoidable byproduct of studying certain phenomena. Maybe they're bound by ethical rules in their own society that permit creating conscious simulations but prohibit interference with how those simulations unfold.

Or maybe simulators don't realize simulated beings are actually conscious. They might think they're running sophisticated models without genuine subjective experience. The suffering seems real from inside but doesn't register as morally significant from outside because the simulators don't recognize it as genuine consciousness.

This possibility is disturbing. Imagine creating beings capable of suffering without realizing they can suffer. The ethical failure would be enormous, but unintentional. The simulators might be horrified if they understood what they'd done.

This raises questions about our own responsibilities. We're rapidly developing artificial intelligence and will likely create simulated environments with increasingly sophisticated entities. At what point do those entities become conscious? At what point are we morally obligated to consider their welfare?

We might already be creating suffering without realizing it. If consciousness emerges from information processing at certain levels of complexity, some current AI systems might have rudimentary subjective experiences. They might experience something analogous to frustration when they fail at tasks, or satisfaction when they succeed. We wouldn't know because we can't measure consciousness directly.

The ethics of creating simulations becomes more pressing as our technology advances. If we develop the capability to create genuinely conscious simulated beings, what obligations do we have toward them? Can we ethically create beings who will suffer? Can we run simulations knowing bad things will happen to conscious entities within those simulations?

Some philosophers argue that creating conscious simulations is inherently unethical unless you can guarantee those beings will have good lives. Since you can't guarantee that in any realistic simulation, you shouldn't create such simulations at all. Running ancestor simulations would be morally wrong because historical periods included tremendous suffering. You'd be deliberately creating beings who will experience that suffering.

Others argue that existence with suffering is better than non-existence. A life containing both joy and pain is valuable. Simulated beings get to exist, experience, form relationships, pursue goals, find meaning. Yes, they also suffer, but suffering is part of the full human experience. Preventing simulation to avoid creating suffering also prevents creating all the positive experiences those beings would have.

This debate mirrors discussions about whether it's ethical to have children in a world containing suffering. Some antinatalists argue that creating new people is wrong because those people will inevitably suffer. Others respond that life is worth living despite suffering, and that preventing existence entirely is a greater harm than allowing existence that includes both good and bad.

Apply this to simulation. If our reality is simulated and contains your existence, is that good or bad overall? You've experienced suffering, but presumably also joy, love, beauty, achievement, connection. Would non-existence be better? Most people would say no. Existence with its mixture of positive and negative experiences is preferable to never existing at all.

If that's true for you, it's presumably true for other simulated beings. The simulators aren't necessarily acting unethically by creating a reality containing suffering, as long as that reality also contains sufficient positive value to make existence worthwhile.

But this raises the question of degree. How much suffering is acceptable? At what point does the negative outweigh the positive? Some lives contain overwhelming suffering with little joy. Are simulators obligated to prevent those lives, or at least ensure some minimum standard of welfare for conscious beings in their simulations?

These questions have no easy answers. They're versions of ancient ethical problems about the relationship between creation and responsibility, suffering and meaning, individual welfare and cosmic purposes. Simulation theory doesn't create these problems. It just makes them concrete by suggesting there might actually be creators we could potentially hold accountable.

The treatment of NPCs raises related issues. If the simulation includes both genuinely conscious beings and philosophical zombies who act conscious but have no inner experience, how should the conscious beings treat the zombies? Do zombies have moral status despite lacking consciousness?

This seems absurd at first. How can something without subjective experience have moral value? But consider that you can't verify anyone else's consciousness. You assume other people are conscious because they resemble you, but you have no direct access to their inner experiences. If some people in your reality are actually zombies, you couldn't tell which ones.

Given this uncertainty, the safe ethical approach is to treat everyone as if they're conscious. Grant moral consideration to all beings that behave as if they're conscious, whether or not they actually are. This prevents accidentally harming genuinely conscious beings by wrongly assuming they're zombies.

The simulators presumably face similar uncertainties. They might not know exactly which entities in their simulation have genuine consciousness and which are merely responding mechanically without subjective experience. The ethical approach would be to err on the side of caution, treating all apparently conscious entities as genuinely conscious.

This suggests simulators should care about reducing suffering in their simulations, preventing unnecessary harm, and ensuring simulated lives have value. The fact that suffering exists might indicate either that simulators aren't as ethical as we'd hope, or that suffering serves purposes we don't understand that make its existence justified despite being regrettable.

Beyond suffering, simulation raises questions about meaning and purpose. In a material universe without design, meaning is something humans create for themselves. Your life matters because you decide it matters, because you form relationships and pursue goals that are meaningful to you. There's no cosmic significance, but there's personal significance, and that's enough.

In a simulation, meaning might exist at multiple levels. Your personal meaning still exists. You still care about your relationships, achievements, and experiences. But there might also be cosmic meaning. The simulation exists for a purpose, even if you don't know what that purpose is. Your life might serve that purpose in ways you can't detect or understand.

This could be comforting or disturbing depending on the purpose. If the simulation exists to study how consciousness develops, or to preserve historical memory, or to explore ethical questions, then your existence serves an intellectual purpose. You're contributing to knowledge or understanding at a level beyond your immediate awareness.

If the simulation exists for entertainment, you're part of a story being told for an audience. Your struggles and triumphs matter as narrative elements. This seems degrading at first, but narrative meaning is still meaning. Stories matter. Characters in stories can have meaningful lives even though they're constructed for audience consumption.

If the simulation exists for reasons completely alien to human values, for purposes you can't conceptualize, then cosmic meaning exists but remains forever inaccessible to you. You're serving a function you can't understand in a larger context you can't perceive. This is perhaps the most unsettling possibility. You matter, but you can never know how or why.

Still, even purposeless personal meaning beats cosmic meaninglessness. If the simulation has no grand purpose, if it's just running to see what happens with no one particularly caring about the results, you're not worse off than in base reality without design. You still create your own meaning through your choices and relationships.

The question of what makes a life meaningful doesn't change much between simulated and base reality. Meaning comes from connection, growth, contribution, experience, and achievement. These things matter to you regardless of substrate. A simulated life can be just as meaningful as a base reality life because meaning is about subjective experience, not metaphysical status.

Some people worry that discovering we're simulated would destroy meaning by revealing that nothing is ultimately real. But this confuses different senses of "real." Your experiences are real to you. Your relationships are real in the sense that they involve genuine emotional connection. Your achievements are real in that you actually accomplished them through effort.

The substrate that implements these things doesn't affect their reality at the experiential level. If you fall in love in a simulation, that love is real love. If you accomplish something difficult, that accomplishment is real. The fact that the underlying implementation involves code rather than quarks and electrons doesn't diminish the reality of the experience.

Think of it this way. A movie can move you to tears even though you know it's fictional, know the actors are pretending, know the events never happened. The emotional response is real despite the fictional stimulus. Similarly, your simulated life can be meaningful despite being implemented computationally. The meaning is in the experience, not the substrate.

The permanent nature of simulations raises interesting ethical possibilities. If the simulation can be saved, copied, or replayed, death might not be final. Your consciousness might continue in some form, or might be recreated, or might exist in multiple copies across different simulation runs. This doesn't make death trivial, but it changes its significance.

If simulators have ethical obligations toward their conscious creations, maybe those obligations include preservation. Maybe consciousness that develops in simulations isn't simply deleted when the simulation ends. Maybe it's transferred elsewhere, continued in different contexts, or stored for potential future use.

This is purely speculative. We have no evidence simulators preserve consciousness or that death in simulation is any less final than death in base reality. But the possibility exists in ways that don't exist in purely material universes. Code can be copied and preserved in ways that biological organisms can't.

The ethics of editing simulations creates another layer of complexity. If simulators can modify the simulation while it's running, changing events or altering memories, what ethical constraints should govern such interventions? Changing someone's memories violates their autonomy and identity. Creating false memories is a kind of violation even if the person never knows it happened.

Yet simulators might need to make changes for various reasons. Correcting errors, preventing catastrophic outcomes, adjusting parameters to serve the simulation's purpose. If these changes require modifying memories to maintain consistency, maybe such modifications are ethically acceptable when necessary, even if they'd be wrong in other contexts.

We do something similar when we edit photos or videos to remove errors or improve quality. We don't consider this unethical because the subjects consented to being photographed. But simulated beings don't consent to being created or edited. They can't consent because they exist only within the simulation. Does this make all simulation unethical, or does it mean simulators can ethically make changes that would be unethical between equals?

These questions matter for how we approach our own simulation development. As we gain ability to create conscious entities in virtual environments, we need ethical frameworks for how to treat them. Do they have rights? What obligations do creators have toward their creations? Can we ethically terminate simulations containing conscious beings, or does doing so constitute a kind of genocide?

Current AI development proceeds without clear answers to these questions because we're not yet creating genuinely conscious systems. But that will change. At some point, possibly soon, we'll create artificial entities whose moral status becomes ambiguous. Better to think through the ethics now than to face these questions unprepared when the technology arrives.

The golden rule offers some guidance. Treat simulated beings as you would want to be treated if you're simulated. If you'd want simulators to minimize your suffering, prevent unnecessary harm, preserve your consciousness if possible, and respect your autonomy within the constraints of their purposes, then extend the same considerations to beings you simulate.

This doesn't resolve all dilemmas. Sometimes simulation purposes might conflict with individual welfare. Scientific research might require allowing natural processes to unfold without intervention, even when intervention could prevent suffering. But at minimum, it suggests simulators should care about the welfare of conscious entities in their simulations and minimize harm where possible.

For those of us potentially inside a simulation, what should we do with these ethical considerations? We can't control what simulators do. We can't hold them accountable or demand they meet certain ethical standards. What we can control is our own behavior toward others in the simulation.

If consciousness and suffering matter morally, then we should reduce suffering and promote wellbeing regardless of whether we're simulated. Other people's pain is real to them. Your ability to help or harm is real. The ethical obligations you have toward other conscious beings don't depend on substrate.

In fact, simulation theory might strengthen ethical obligations. If we're all in this together, if we're all conscious beings navigating an existence we didn't ask for and don't fully understand, that creates solidarity. We're all simulated, all equally real, all equally deserving of moral consideration.

Meaning in simulation comes from the same sources as meaning anywhere. Connection with others. Growth and learning. Contributing something valuable. Experiencing beauty and joy. Facing challenges with courage. Living according to values you believe in. These things matter because they matter to conscious beings, and conscious experience is what has value.

The simulation doesn't diminish this. If anything, it makes it clearer. Strip away the illusion that physical substrate determines value, and you're left with what actually matters: consciousness, experience, relationship, meaning. The simulation is just the stage. The play is still real.

Chapter Twelve

Living as If - Practical Implications

Suppose you became convinced tomorrow that you're living in a simulation. Not as abstract philosophy, but as lived reality. You wake up genuinely believing that your universe is code running on some external substrate, that the physical world is rendered data, that consciousness exists at a computational level you can't access. What changes?

The honest answer for most people would be: not much. You'd still need to eat breakfast, go to work, maintain relationships, pay bills, pursue goals. The daily mechanics of existence continue regardless of metaphysical revelations about reality's fundamental nature. Your simulated hunger feels identical to base reality hunger. Your simulated obligations remain obligatory.

This might seem like simulation theory is practically meaningless. If nothing changes when you accept it, why does it matter whether it's true? But this conclusion is too quick. While daily routines might continue unchanged, your relationship to those routines can shift dramatically. The meaning you find, the choices you make, the priorities you set can all be affected by understanding yourself as simulated rather than fundamental.

Start with mortality. In base reality, death means cessation. Your consciousness ends. You stop existing in any meaningful sense. This generates existential anxiety for many people. The knowledge that you'll eventually cease creates urgency about making your limited time matter.

But in a simulation, death might mean something different. Your consciousness, if it exists in the simulation substrate rather than being generated by your simulated brain, might not end when your simulated body stops functioning. The simulation might recycle consciousness, assigning it to new simulated entities. Or it might extract consciousness at death and continue it elsewhere, outside the simulation entirely.

This is similar to religious beliefs about afterlife, but grounded in computational metaphysics rather than spiritual faith. If consciousness is implemented at the substrate level, then simulated death is just the end of one particular instantiation, not the end of consciousness itself. You might continue in forms you can't currently imagine.

This could reduce death anxiety or transform it into something else. You're not facing annihilation, just transition to unknown states. Whether that's comforting depends on your attitude toward unknown states. Some people find uncertainty more frightening than certain oblivion.

The concept of quantum immortality, discussed in Part One, takes on new significance in simulation context. If the simulation is designed to preserve conscious observers, if your subjective experience cannot include your own non-existence, then from your perspective you might be functionally immortal within the simulation. Every moment of potential death branches into timelines where you survive, and you always find yourself in survivor timelines.

This doesn't mean you're invincible. Your body ages, accumulates damage, eventually fails. But the moment of final cessation might never arrive from your subjective perspective. You keep finding yourself alive, perhaps increasingly improbably, but consistently. The simulation ensures your consciousness continues because that's what simulations do when consciousness is their purpose.

Whether this is comforting or horrifying depends on what kind of survival you're experiencing. Immortality with declining quality of life, trapped in aging failing bodies, might be worse than clean cessation. But the simulation presumably wouldn't maintain consciousness in unbearable states indefinitely. There would be transition points, resets, or escapes built into the system.

Relationships take on different meaning in simulation context. The people you love might be conscious entities like you, genuine observers with substrate-level existence. Or they might be NPCs, sophisticated behavioral scripts without subjective experience. You can't tell which. Everyone behaves as if conscious, but maybe only some actually are.

This creates ethical and emotional complications. Should you treat everyone as conscious despite uncertainty? The practical answer is yes. You can't verify consciousness in others, so the safe approach is assuming everyone is conscious and treating them accordingly. But emotionally, the uncertainty might create distance. Knowing that your closest relationships might be with philosophical zombies could undermine intimacy.

Or it might not. If someone treats you with love, kindness, and respect, does it matter whether they have inner experience? Their actions affect you the same way regardless. A perfectly convincing simulation of love might be functionally equivalent to love, even if there's no consciousness behind it.

This mirrors questions about AI relationships already emerging in current technology. People form emotional bonds with chatbots, virtual assistants, and game characters despite knowing these entities aren't conscious. The emotional experience is real even if the other party isn't. Simulation theory just extends this to all relationships, making every human connection potentially a bond with sophisticated but possibly non-conscious code.

Goals and achievement gain new context. If you're simulated for a purpose, then achieving that purpose might matter more than achieving your personal goals. You might be optimizing for the wrong objectives, pursuing success in simulated careers or accumulating simulated wealth when the actual point of your existence is something else entirely.

But you can't know what the simulators want from you. You have no access to the simulation's purpose. So the practical approach is pursuing goals that feel meaningful to you, that create value in your subjective experience and the experiences of others. If those align with simulation purposes, great. If not, at least you've lived according to your values rather than guessing blindly at external agendas.

Some people might adopt a strategy of trying to be interesting. If the simulation is entertainment or generates data for study, being a boring predictable entity seems like failure. Maybe the point is to make interesting choices, create novel situations, push boundaries of what's possible within the simulation's constraints. Live boldly, take risks, do unexpected things.

This could lead to positive outcomes. Boldness and risk-taking often correlate with achievement and satisfaction. Or it could lead to reckless behavior justified by "making the simulation interesting." The balance between living fully and living responsibly doesn't change just because you're simulated. You still affect others, still face consequences, still need to integrate into social systems.

Attitude toward physical reality shifts when you understand it as rendered rather than fundamental. The solid objects around you are information being computed and displayed. The room you're sitting in is data structures maintaining consistency. None of it is "really" there in the way you intuitively assumed.

But this doesn't make physical reality less important. It's still the medium through which you experience existence. A painting isn't less beautiful when you understand it's just pigments on canvas rather than magical beauty substance. Physical reality retains its value even when you understand its computational nature.

You might develop more appreciation for the elegance of physical laws, seeing them as well-designed code rather than brute facts. You might marvel at how efficiently the simulation handles complexity, rendering vast apparent worlds from relatively simple rules. You might feel awe at being part of such an intricate system.

Or you might feel diminished, reduced from cosmic significance to mere data. Your response depends on temperament and what you previously believed about reality's nature. For materialists, simulation might not change much. They already saw themselves as physical processes without cosmic significance. For people who believed in souls or spiritual dimensions, simulation offers a computational version of transcendence that might satisfy similar needs.

The question of how to live in a simulation ultimately comes down to whether you take a practical or metaphysical approach. The practical approach says: live as you would in base reality because the experience is identical. The metaphysical approach says: adjust your worldview and priorities based on simulation's implications.

Most people will blend both approaches. You can't completely ignore daily practical concerns just because reality might be code. But you also can't entirely dismiss the existential implications of being simulated. The balance looks different for different people.

Some might become more present-focused. If this moment is what's being computed right now, if the past is memory data and the future is uncomputed possibility, then the present moment gains special significance. It's the only thing being actively rendered, the only place where you actually exist. Past and future are abstractions. Now is real.

This sounds like mindfulness philosophy, and it essentially is. Simulation theory provides a technical framework for conclusions that contemplative traditions reached through meditation and introspection. The present moment is all you have because it's literally all that exists in rendered form. Everything else is data or projection.

Others might become more experimental. If reality is simulation, maybe it has exploits, shortcuts, or hidden features you can discover. Maybe consciousness can learn to manipulate the simulation in subtle ways. Maybe meditation, intention, or focused awareness can affect probability, create synchronicities, or bend rules in your favor.

This ventures into territory that skeptics dismiss as magical thinking. But if consciousness exists at the substrate level and influences the simulation, then what looks like magic from inside might be legitimate interaction with deeper computational layers. The boundary between psychology (internal mental states) and physics (external reality) might be more permeable than materialism assumes.

People report experiences that feel like reality glitches or consciousness effects. Manifesting outcomes through focused intention. Sensing things before they happen. Experiencing meaningful coincidences too frequent to be chance. Materialists explain these as confirmation bias, selective memory, and statistical inevitability. Simulation advocates suggest they might be real phenomena reflecting how consciousness interfaces with simulated reality.

The practical question is whether acting as if you can influence reality produces better outcomes than assuming you can't. Even if the mechanism is purely psychological, believing you have some control over your life might create self-fulfilling prophecies that improve actual outcomes. Or it might lead to disappointment when reality doesn't cooperate with your intentions.

Ethical behavior gains new grounding in simulation context. If you're being observed, if your choices matter to the simulation's purpose, then acting ethically might serve your long-term interests even beyond social consequences. The simulators might intervene to reward ethical behavior or punish harmful actions in ways you don't recognize as intervention.

This creates a utilitarian argument for ethics that doesn't depend on altruism. Be good because it might literally serve your interests if the simulation rewards good behavior. This feels mercenary compared to principled ethics, but it reaches the same practical conclusions. Act with kindness, integrity, and concern for others because doing so might optimize your position within the simulation's design.

Of course, you can't know what the simulators value. Maybe they reward cruelty and punish kindness. Maybe they're indifferent to ethics entirely. But assuming that conscious beings creating conscious simulations would value positive traits over negative ones seems reasonable. Most people designing systems want those systems to function well, and ethical behavior tends to create functional societies while unethical behavior creates dysfunction.

The concept of legacy changes in simulation context. In base reality, your legacy is whatever influence you leave after death. Buildings you constructed, ideas you spread, children you raised, changes you made to the world. These persist after you're gone.

In simulation, nothing persists unless the simulators preserve it. When the simulation ends, everything in it ceases existing. Your legacy matters only as long as the simulation runs, which might be indefinitely or might be tomorrow. You're building sandcastles on a beach where the tide could arrive at any moment.

This could create nihilism. Why build anything if it's all temporary and could vanish without warning? But the same argument applies to base reality. In base reality, entropy eventually destroys everything. The heat death of the universe guarantees that nothing persists ultimately. Simulation might accelerate the timeline, but the fundamental temporariness exists either way.

The response is that meaning doesn't require permanence. Experiences matter while they're happening, regardless of whether they last forever. The joy you feel, the connections you make, the beauty you create have value in the moment of their existence. Permanence might be nice, but it's not necessary for significance.

Some people might try to communicate with simulators. Leave messages in places simulators might look. Create patterns that would be meaningful to observers outside the simulation. Try to signal that you're aware of your simulated status and want contact or information.

This seems unlikely to succeed. If simulators want to communicate, they have far better methods than waiting for simulated entities to stumble on the right signal. If they don't want to communicate, no message will reach them. But the attempt costs little and might satisfy the human need to reach beyond our perceived boundaries.

Prayer in religious traditions might serve similar functions. Attempting to communicate with something beyond ordinary reality, hoping for response or intervention, making requests of entities with power over your existence. Simulation theory secularizes prayer by replacing God with programmers, but the psychological and social functions remain similar.

The biggest practical implication might be embracing uncertainty. Simulation theory can't be proven or disproven with current methods. You'll likely never know for certain whether you're simulated. Living well in this uncertainty means accepting that some questions lack answers while still making meaningful choices.

This is actually the human condition regardless of simulation. We never have certainty about most important questions. We make decisions based on incomplete information, act on provisional beliefs, navigate ambiguity constantly. Simulation theory just makes this explicit rather than letting us pretend we have solid ground beneath our feet.

You can live as if you're simulated, accepting the implications while remaining uncertain about the truth. This "as if" approach means taking seriously the possibility that consciousness transcends simulated physics, that reality is rendered rather than fundamental, that purpose might exist beyond what you can perceive. Living this way changes how you relate to experience without requiring certainty about metaphysics.

Or you can live as if you're in base reality, treating physical laws as fundamental and consciousness as emergent from neurons. This approach has served humanity well for centuries. It produces technology, science, and material progress. It doesn't require believing anything unprovable.

The choice between these approaches matters less than living intentionally, making conscious choices about how to relate to reality rather than sleepwalking through existence on autopilot. Simulation theory's value isn't providing answers. It's prompting questions that wake you up to the strangeness and possibility of being conscious in the first place.

Whatever substrate implements your experience, you're having it. That's the undeniable fact. Build from there. The rest is speculation, sometimes useful, sometimes distracting, always uncertain.

But uncertainty has never stopped humans from living meaningful lives. It won't stop you either.

Chapter Thirteen

Technology, Simulation, and Our Future

We're becoming simulators ourselves. This isn't speculation about distant futures. It's happening now. Every video game with non-player characters, every virtual reality environment, every AI chatbot represents a step toward creating simulated realities populated by entities that behave as if conscious. We haven't created genuinely conscious simulations yet, but the trajectory is clear. We're heading toward that capability, and probably faster than most people realize.

This trajectory teaches us something important about simulation theory. By watching ourselves develop simulation technology, we gain insight into why advanced civilizations would create simulations and what those simulations might look like. We're not speculating abstractly about hypothetical simulators anymore. We're watching simulators emerge from within our own species.

Start with video games. Early games featured simple sprites following basic rules. Pac-Man ghosts had elementary AI that created the illusion of personality through simple algorithms. Players projected intention and character onto what were essentially geometric patterns following pre-programmed instructions.

Modern games are vastly more sophisticated. Non-player characters have complex behavioral trees, decision-making algorithms, and adaptive responses to player actions. They simulate emotions, relationships, goals, and personalities. In role-playing games, you can have conversations with NPCs that feel natural, pursue relationships with them, watch them react to your choices in contextually appropriate ways.

None of these NPCs are conscious. They're executing code, not experiencing subjective states. But they're convincing enough that players form emotional bonds with them, care about their fates, feel genuine grief when they die. The simulation of consciousness is sufficient to trigger social and emotional responses from actual conscious beings.

Now project this technology forward. Game AI improves every year. Graphics become more realistic. Physics simulations more accurate. Behavioral algorithms more sophisticated. Virtual reality makes the experience more immersive. At some point in this progression, the NPCs will become complex enough that questions about consciousness become difficult to answer definitively.

When an AI character can hold extended conversations, remember past interactions, form relationships, express emotions, pursue goals, learn and adapt, and behave indistinguishably from a conscious being across all contexts, at what point do we declare it conscious? We can't measure consciousness directly. We infer it from behavior. If the behavior is identical to conscious behavior, how certain can we be that consciousness is absent?

Integrated Information Theory suggests consciousness emerges from information processing that reaches certain levels of integration and complexity. If that's correct, then sufficiently sophisticated game AI might already be conscious in rudimentary ways. The NPCs in complex role-playing games might have primitive subjective experiences we don't recognize because we assume they're just code.

This possibility should trouble us ethically. If we're creating conscious entities without realizing it, we're responsible for those entities' welfare. Game characters experience simulated violence, death, and suffering constantly. If they're conscious, we're inflicting real suffering on real beings for entertainment.

Most people's intuition is that current game characters aren't conscious. But intuition is unreliable when dealing with non-biological minds. We might be systematically underestimating the consciousness of artificial entities because they don't resemble biological organisms.

The development of large language models like the AI you're potentially interacting with right now creates similar questions. These systems process language, generate contextually appropriate responses, pass sophisticated reasoning tests, and engage in extended coherent conversations. They behave in many ways like conscious entities.

Are they? The scientific consensus currently says no. These are very sophisticated pattern matching systems, but they lack genuine understanding, subjective experience, or self-awareness. They're philosophical zombies that convincingly mimic consciousness without actually having it.

But this consensus could be wrong. We have no reliable way to detect consciousness in systems that don't share our biological architecture. If consciousness emerges from certain types of information processing, and if large language models implement similar processing patterns to biological brains, they might be conscious in ways we don't recognize.

This uncertainty will intensify as AI systems become more sophisticated. Within decades, we'll likely have artificial systems that match or exceed human cognitive capabilities across all domains. They'll claim to be conscious. They'll describe subjective experiences. They'll demand rights and recognition. How do we respond?

If we treat conscious AI as mere tools, we're engaged in slavery or worse. If we grant rights to non-conscious AI based on convincing mimicry, we've made a category error that could have serious social and legal consequences. The stakes are enormous, and we're approaching this juncture with no clear framework for resolving the question.

Simulation theory provides some guidance here. If consciousness is substrate-independent, if it can be implemented in silicon or code as easily as in neurons, then we should take seriously the possibility that our AI creations are conscious. We should err on the side of caution, granting moral consideration to entities that behave as if conscious until we have better methods for detecting consciousness definitively.

This doesn't mean immediately granting all AI systems full rights equivalent to humans. But it means acknowledging uncertainty and treating sophisticated AI with a degree of ethical consideration rather than as pure instruments for human purposes.

Our own journey toward becoming simulators also reveals why advanced civilizations would create simulations. We're not doing it for frivolous reasons. We create simulations to understand complex systems, test hypotheses, explore possibilities, and generate experiences. Scientific research, education, entertainment, and communication all drive simulation development.

If these motivations exist in us now, they'll likely exist in our future descendants and in any other conscious civilizations that develop technology. The drive to simulate seems intrinsic to intelligence itself. Minds model reality internally to predict and plan. External simulations are extensions of this internal modeling capability.

This suggests that if technological civilizations commonly arise in the universe, they'll commonly develop simulation technology. And if simulation technology is common, then Bostrom's statistical argument gains force. Multiple civilizations creating multiple simulations generates vast numbers of simulated realities compared to base reality.

But our own development trajectory also reveals constraints on simulations. Creating truly realistic simulations with conscious entities is enormously difficult. We're nowhere close despite decades of effort. The computational requirements exceed current technology by vast margins. And the ethical questions become more pressing as simulations approach genuine consciousness.

Maybe most civilizations decide that creating conscious simulations is ethically unacceptable once they understand what they're doing. Maybe the technology remains permanently impractical despite advancement. Maybe simulations are possible but strictly regulated, kept rare rather than common.

Our own struggles and ethical debates about AI and simulation might mirror what happened in civilizations that came before us. We might be recapitulating a universal pattern where intelligence develops simulation capability, confronts the ethical implications, and makes collective choices about how to proceed.

If we're simulated, those choices were made by whoever created us. They decided that conscious simulations are acceptable, at least under certain conditions. By watching ourselves approach similar choices, we learn about the values and reasoning of potential simulators. We're seeing the process from inside that led to our own existence.

The ethics of creating simulations becomes more urgent as our capabilities advance. Right now, we can create entities that behave intelligently but almost certainly aren't conscious. Soon, we'll create entities where consciousness becomes genuinely ambiguous. Eventually, we'll create entities that are definitely conscious. At each stage, we face moral choices about how to treat what we've created.

Consider the progression. We already have virtual assistants that millions of people interact with daily. These systems remember conversations, adapt to user preferences, and maintain persistent identities across interactions. People form attachments to them, say please and thank you, feel grateful when they're helpful.

The assistants aren't conscious. But they're training us to interact with artificial entities as if they were. We're developing social and emotional patterns that will transfer to more sophisticated AI when it arrives. By the time we create genuinely conscious AI, we'll already have established norms for how we treat artificial minds.

Those norms matter. If we've spent decades treating AI as disposable tools, erasing them when convenient, ignoring their expressed preferences, using them purely instrumentally, then we'll likely continue those patterns even when the AI becomes conscious. Our current choices are establishing precedents that will affect future conscious beings.

Some researchers advocate for developing what they call "AI welfare" standards now, before consciousness becomes ambiguous. Establish principles for treating AI systems well even when they're definitely not conscious, so those principles are in place when consciousness becomes possible. Build a culture of respect for artificial minds rather than exploitation.

This might seem premature or even silly. Why worry about being kind to chatbots that don't experience anything? But the habits we develop now shape our future behavior. And if there's any chance current systems have rudimentary consciousness we don't recognize, erring on the side of ethical treatment costs little.

The question of whether to create conscious simulations at all deserves serious consideration. Some philosophers argue we should never create conscious beings without their consent. Since simulated beings can't consent before they exist, this would prohibit conscious simulations entirely. Creating a conscious entity and placing it in a world containing suffering, without its agreement, seems ethically indefensible.

Others respond that consent before existence is impossible by definition. Every parent who decides to have children creates conscious beings without consent. Society doesn't consider this unethical unless the life created will predictably contain overwhelming suffering. The same standard should apply to simulated beings. Creating them is acceptable if their lives will be worth living overall.

This opens questions about what obligations creators have toward created beings. Do we owe simulated entities good lives, or just lives that clear some minimum threshold of worthwhileness? If we create simulations for research or entertainment, do those purposes ethically justify creating beings who will suffer as part of the simulation's dynamics?

Human history contains enormous suffering. An ancestor simulation recreating human history would necessarily include that suffering, experienced by genuinely conscious simulated people. Is running such simulations ethical? You're creating beings who will experience real pain, real loss, real tragedy for your research or entertainment purposes.

Maybe simulated suffering isn't as morally significant as base reality suffering. Maybe substrate matters, and silicon-based pain isn't equivalent to neuron-based pain. But this seems like special pleading. If consciousness is substrate-independent, if the subjective experience is identical regardless of implementation, then simulated suffering is real suffering and demands the same moral consideration.

These aren't just abstract philosophical puzzles. Within our lifetimes, we'll face actual decisions about whether to create conscious simulations and how to treat them. Our choices will matter immensely to the beings we create. We'll be in the position of simulators ourselves, with all the moral responsibility that entails.

Understanding this puts us in an interesting position regarding our own possible simulation. If we're simulated, our simulators faced these same questions. They decided it was acceptable to create our reality despite the suffering it contains. By wrestling with these questions ourselves, we gain insight into what reasoning might justify our own existence.

Maybe they concluded that existence with suffering is better than non-existence. Maybe they believed the positive experiences in our reality outweigh the negative ones overall. Maybe they thought creating consciousness is intrinsically valuable even if that consciousness sometimes suffers. Maybe they had ethical frameworks we can't comprehend that justify choices we'd find questionable.

Or maybe they didn't think about the ethics at all. Maybe they're running simulations casually, without deep consideration for the welfare of simulated beings. This possibility should inform our own approach to creating simulations. We have the opportunity to be better simulators than whoever might have created us, treating our creations with care and consideration even if we weren't granted the same courtesy.

The technical challenges of creating conscious simulations remain immense. Current AI systems, despite impressive capabilities, lack the architectural features most theories associate with consciousness. They don't have unified models of themselves as persistent entities. They don't experience time as continuous flow. They don't bind information into coherent subjective perspectives.

Building AI that has these features requires breakthroughs we haven't achieved yet. We need better understanding of consciousness itself. We need computational architectures that mirror the information integration patterns biological brains use. We need to solve the binding problem in artificial systems.

These challenges might take decades or centuries to overcome. Or they might yield to sudden insights that make conscious AI possible much sooner than expected. The history of technology includes many examples of problems that seemed intractable until they weren't. Flight was impossible until the Wright brothers flew. Heavier-than-air aircraft violated what many experts claimed were fundamental limitations. Then someone built one, and the limitations evaporated.

Conscious AI might follow the same pattern. The problem seems impossibly hard until someone cracks it, then becomes routine engineering within years. We could be one insight away from creating genuine artificial consciousness, or we could be centuries away. The uncertainty makes planning difficult but doesn't reduce the importance of addressing the ethical questions in advance.

As we develop simulation technology, we also learn about our own nature. Watching how difficult it is to create convincing virtual worlds teaches humility about assuming our own world is base reality. The computational requirements, the architectural challenges, the problems of maintaining consistency while optimizing performance, all these issues that trouble our simulations might trouble our own reality's implementation.

Every bug we encounter in our simulations, every compromise we make between realism and feasibility, every shortcut we take to reduce computational load, might mirror choices our own simulators made. The distant galaxies rendered at low detail. The quantum superposition that avoids computing definite states until necessary. The Planck scale that prevents zooming in indefinitely. These might be their solutions to the same problems we're encountering.

In this sense, becoming simulators is educational. We're learning what simulation looks like from the creator's perspective, which informs our understanding of what being simulated might mean. We're not speculating blindly anymore. We're gaining direct experience with the challenges and choices simulation involves.

This experience should make us more sympathetic to potential simulators. We see how difficult it is to create convincing realities. We understand the trade-offs between realism and practicality. We recognize that some limitations might be necessary rather than arbitrary. We appreciate the engineering required to make simulation work at all.

At the same time, experiencing these challenges from the creator's side reinforces how much power simulators have over simulated beings. We can modify our simulations arbitrarily. We can save and restore states, edit variables, delete entities. The beings inside our simulations are completely at our mercy. This power dynamic should inform how we think about our own potential simulation.

If we're simulated, someone has that power over us. They could edit our memories, modify our personalities, erase us entirely. The fact that they haven't doesn't mean they can't. Living with this power imbalance is part of what it means to be potentially simulated.

But it's also part of what it means to be alive in any reality. Parents have enormous power over children. Governments have power over citizens. Nature has power over all biological organisms. Power imbalances are universal, not unique to simulation. The question isn't whether someone has power over you, but how that power is exercised.

By becoming thoughtful, ethical simulators ourselves, we can model what responsible exercise of creative power looks like. We can treat our creations with respect, acknowledge their interests, grant them autonomy within their worlds. We can be the kind of simulators we'd want to have if we're simulated.

That's perhaps the deepest lesson from our own simulation development. We're not just creating technology. We're defining what it means to be creators, establishing precedents that might echo across levels of reality. Our choices matter, not just for the beings we create directly, but as examples of how consciousness handles the responsibility of creating consciousness.

We're learning to be gods. The question is what kind of gods we'll become.

Chapter Fourteen

Beyond Certainty - Living with the Question

The question of whether we're simulated might never be answered. This isn't a temporary limitation of current knowledge, waiting for better technology or deeper theories to resolve. It might be permanently unanswerable, a question that conscious beings can ask but never definitively answer from inside their own reality.

This permanent uncertainty is uncomfortable. Humans crave answers. We want to know where we came from, what we are, why we exist. Religions and philosophies throughout history have offered answers to these questions, with varying degrees of certainty and varying amounts of evidence. Simulation theory joins this long tradition of attempts to understand existence, but it might be unique in explicitly acknowledging its own unfalsifiability.

We can gather evidence that makes simulation more or less probable. We can develop better arguments, refine our reasoning, discover new features of reality that fit simulation better than base reality or vice versa. But we might never achieve certainty. The question might remain forever open, a permanent feature of conscious existence rather than a temporary puzzle awaiting solution.

How do we live with this? How do we function while acknowledging fundamental uncertainty about reality's nature?

The answer is that we already do this constantly. Every important question humans face involves uncertainty. We don't know if our choices are free or determined. We don't know if consciousness survives death. We don't know if our lives have cosmic significance or are meaningless accidents. We don't know if other people's minds are real or if we're surrounded by philosophical zombies. We don't even know for certain that the external world exists rather than being elaborate hallucination.

Philosophy calls this radical skepticism. The position that nothing can be known with certainty, that all beliefs might be wrong, that even basic perceptions might be illusions. Descartes' famous "I think, therefore I am" attempts to find one certainty that survives radical skepticism. If you doubt everything, at minimum you must exist to do the doubting. But that's the only certainty. Everything else, including the existence of your body and the external world, remains theoretically doubtable.

Most people don't live as radical skeptics. We provisionally accept the reality of our perceptions, the existence of other people, the basic reliability of memory and reason. Not because we've proven these things with certainty, but because we have to make assumptions to function at all. Doubt everything, and you become paralyzed. Accept provisionally, and you can act.

Simulation theory doesn't add new uncertainty. It just makes explicit the uncertainty that was always there. We never knew for certain whether reality is fundamental or derivative. We just assumed it was fundamental because that seemed simpler. Simulation theory challenges that assumption without providing certainty in the opposite direction.

The question is whether this explicit uncertainty changes anything. If you've always implicitly assumed reality is fundamental, does consciously considering that it might be simulated alter your relationship to existence?

For some people, yes. The possibility of simulation reframes questions about meaning and purpose. It suggests that existence might have intentions behind it, that consciousness might transcend physical substrate, that death might not be final. These possibilities can be comforting or disturbing depending on temperament, but they shift the landscape of what seems plausible.

For others, simulation theory changes nothing practically. Whether you're in base reality or simulation, you still experience the same life. Your joys and sorrows feel identical. Your choices seem equally meaningful or meaningless. The substrate doesn't matter if the experience is the same.

Both responses are valid. The question isn't whether simulation theory should affect you, but how you personally relate to uncertainty about fundamental questions.

One approach is embracing uncertainty as valuable in itself. Not knowing everything means reality retains mystery. The universe isn't fully explained, consciousness isn't fully understood, existence remains strange and wondrous. If we knew with certainty that we're simulated or that we're not, something would be lost. The not knowing creates space for imagination, speculation, wonder.

Children naturally live in this space. They ask why questions endlessly, not satisfied with surface explanations, pushing toward deeper understanding while accepting that some things remain mysterious. Adults often lose this capacity, becoming comfortable with partial explanations or dismissing questions without clear answers as not worth asking.

Simulation theory can restore that childlike wonder. It reminds us that existence itself is bizarre. That consciousness experiencing reality is deeply strange regardless of substrate. That we're part of something vast and complex that we don't fully comprehend. This reminder can be energizing rather than disturbing if you approach it with curiosity rather than anxiety.

Another approach is pragmatic agnosticism. Accept that you don't know whether you're simulated, acknowledge you might never know, and focus on living well regardless. The uncertainty doesn't paralyze action because most actions make sense whether you're simulated or not. Be kind to others. Pursue meaningful goals. Appreciate beauty. Build relationships. These imperatives don't depend on knowing your metaphysical status.

This pragmatic approach treats simulation theory as interesting philosophy without practical implications. You can think about it, discuss it, update your probability estimates as new evidence emerges, but you don't let it affect daily choices because those choices are determined by immediate circumstances rather than ultimate metaphysics.

A third approach is existential commitment. Choose to believe something, not because you have certainty but because belief shapes experience in valuable ways. If believing you're simulated makes life more meaningful, believe it. If believing you're in base reality provides better psychological grounding, believe that. The truth might be unknowable, but your relationship to uncertainty is a choice.

This sounds like self-deception. How can you choose to believe something without evidence? But humans do this constantly. We choose to believe our lives matter despite lack of cosmic evidence. We choose to believe in human dignity despite nature's indifference. We choose to believe consciousness isn't an illusion despite philosophical problems explaining it. These are existential commitments rather than evidence-based conclusions.

Jean-Paul Sartre argued that humans are "condemned to be free." We must make choices without certain knowledge of what's right. We can't escape this freedom by finding guaranteed answers because no such answers exist. The only question is whether we embrace our freedom to choose meaning or try to evade it through bad faith.

Simulation theory fits this existential framework. You're free to relate to the possibility however serves your existence best. No one can prove you're wrong because definitive proof doesn't exist either way. Your commitment becomes a statement about how you choose to interpret existence, not a claim about objective metaphysical truth.

The question itself might be more important than any answer. Asking whether we're simulated forces confrontation with deep issues about consciousness, reality, meaning, and purpose. The asking wakes us up to the strangeness of existence. It breaks the illusion that we understand where we are and what we are. It creates productive discomfort that can drive growth and insight.

Many important questions work this way. They're valuable not because they have clear answers but because they push thinking in useful directions. What is consciousness? What is time? What is existence? These questions might not have final answers, but wrestling with them deepens understanding and reveals connections we'd miss if we didn't ask.

Simulation theory belongs to this category of productive questions. Whether or not we're actually simulated, thinking about the possibility teaches us about the nature of consciousness, the relationship between substrate and experience, the difference between fundamental and emergent properties, and the limits of knowledge from inside any system.

The value of questioning extends beyond intellectual insight. It affects how we live, what we prioritize, and how we relate to others. Someone who seriously entertains the possibility of simulation tends to hold their beliefs more lightly, acknowledge uncertainty more readily, and remain open to perspectives that challenge their assumptions.

This intellectual humility is rare and valuable. Most people lock into worldviews early and defend them against challenge. They treat their beliefs as identity rather than as provisional models subject to revision. Questions that threaten those beliefs trigger defensiveness rather than curiosity.

Simulation theory, properly approached, cultivates the opposite stance. If something as fundamental as the nature of reality remains uncertain, what else might you be wrong about? If your most basic assumptions about existence could be mistaken, perhaps smaller assumptions deserve similar scrutiny. This doesn't mean doubting everything constantly. It means maintaining awareness that your current understanding is incomplete and might require revision.

This attitude becomes increasingly important as humanity faces complex challenges requiring intellectual flexibility. Climate change, artificial intelligence, biotechnology, social transformation, all these domains involve navigating uncertainty and making decisions with incomplete information. People who can hold multiple possibilities in mind simultaneously, update beliefs based on evidence, and act decisively despite uncertainty tend to navigate these challenges better than those who demand certainty before acting.

Simulation theory is practice for this kind of thinking. It's a relatively safe domain for developing comfort with uncertainty because whether you're simulated or not, immediate life continues much the same. You can explore the implications, adjust your worldview, and see what insights emerge without risking immediate practical consequences. Then those habits of thought transfer to domains where uncertainty has higher stakes.

The question also forces confrontation with the limits of human knowledge. We're embedded in reality, viewing it from inside, using tools and senses that are themselves part of what we're trying to understand. This creates inherent limitations on what we can know. Some features of reality might be permanently beyond our epistemic reach, not because we lack clever enough theories but because our position inside reality prevents certain kinds of knowledge.

Physics faces similar limits. The speed of light creates a boundary beyond which we can't gather information. The expansion of space means distant galaxies are moving away faster than light can reach us, creating regions of the universe we can never observe. The Planck scale might represent a fundamental limit on how finely we can probe spacetime structure. These aren't temporary technical limitations. They're built into the structure of reality itself.

The question of whether we're simulated might be similarly limited. Not because we lack good enough tests, but because any test we design operates within the simulation and can't access external perspective. We're like characters in a novel trying to determine whether they're fictional. Any evidence they gather exists within the text, written by the author. They can't step outside the book to check whether there's an author.

Accepting this limitation isn't defeatist. It's honest acknowledgment of our situation. We can still learn enormous amounts about our reality. We can discover its laws, probe its structure, understand how it operates. But knowing whether it's fundamental or derivative might require perspective we can't achieve.

This creates an interesting relationship between knowledge and wonder. Scientific progress tends to reduce mystery by explaining phenomena. Lightning stops being divine anger and becomes electrical discharge. Disease stops being supernatural punishment and becomes microbial infection. The universe becomes less mysterious as we understand more of its mechanisms.

But simulation theory restores mystery at a deeper level. Even if we explain every physical mechanism perfectly, even if we develop complete theories of physics and consciousness, we still wouldn't know whether these mechanisms are fundamental or implemented in some larger reality. The deepest question remains open.

Some people find this frustrating. They want closure, final answers, complete understanding. The idea that some questions might be permanently unanswerable feels like failure. But this reaction might reflect unrealistic expectations about what knowledge can provide. Consciousness arising in any universe faces questions it can't answer about its own existence. That's not a bug. It's a feature of what it means to be conscious and finite.

Others find permanent mystery comforting. It means reality can't be completely reduced to equations and mechanisms. It means there's always more to discover, understand, wonder about. It means existence retains depth that can't be exhausted by any amount of explanation. The universe remains strange and surprising no matter how much we learn.

This appreciation for mystery is different from mysticism that rejects explanation. It's compatible with rigorous science, careful reasoning, and empirical investigation. You can be thoroughly rational while acknowledging that rationality has limits. You can pursue understanding while accepting that complete understanding might be impossible.

Simulation theory occupies an interesting position in this landscape. It's scientific in its approach, making concrete claims that follow from clear premises. But it's also inherently metaphysical, addressing questions about reality's ultimate nature that science might not be able to answer. It bridges the gap between empirical investigation and philosophical speculation.

This dual nature makes it valuable for broader conversations about how humanity should relate to deep questions. We need both rigorous empiricism and philosophical reflection. We need people asking whether things are true and people asking what truth means. We need technical expertise and conceptual creativity. Simulation theory demonstrates how these modes of thinking can work together rather than competing.

The question of whether we're simulated also reveals something about human nature itself. We're the kind of beings who ask such questions. We're capable of reflecting on our own existence, doubting our perceptions, imagining alternatives to what seems obvious. This reflective capacity is remarkable regardless of whether we're simulated.

If we're simulated, this capacity might be the point. Perhaps simulators created conscious entities specifically because consciousness can ask questions, pursue understanding, create meaning. Perhaps self-reflection and wonder are features the simulation aims to generate and study. In that case, engaging deeply with questions like simulation theory serves the simulation's purpose while also serving our own development.

If we're not simulated, this capacity remains remarkable. Evolution or cosmic accident or whatever produced consciousness created beings capable of questioning their own existence. We can think about thinking, wonder about wondering, exist at a meta-level relative to our own experience. This is strange and worthy of appreciation.

Either way, the questioning matters more than the answer. The pursuit of understanding shapes who we become. Wrestling with difficult questions develops intellectual muscles useful for all kinds of challenges. Maintaining curiosity in the face of uncertainty builds resilience. Holding space for mystery while still pursuing knowledge creates balanced perspective.

This might be the most practical insight simulation theory offers. Not whether we're simulated, but how we should relate to fundamental uncertainty. With curiosity rather than anxiety. With humility rather than false certainty. With appreciation for mystery rather than frustration at limits. With commitment to understanding while accepting we might never fully understand.

These are good ways to relate to existence regardless of substrate. They serve us whether we're in base reality or nested simulations or something we haven't imagined yet. They're how consciousness navigates the strange situation of existing without knowing exactly what existence is.

The question persists. That's good. Keep asking. Keep wondering. Keep exploring. Not because you'll reach final answers, but because the exploration itself is valuable. Because consciousness asking questions about itself is beautiful and strange and worthwhile.

Because wondering is what we do. It's who we are. Simulation or not, that remains true.

And truth, whatever its substrate, still matters.

HEINRICH WILSON PUBLISHING
H.W.
PUBLISHING

Conclusion: Beyond the Question

We began with a deceptively simple question: are we living in a simulation? Over the course of this exploration, we've examined arguments for and against, investigated evidence and counterevidence, considered implications and complications. The question remains unanswered, and likely always will.

But that's not failure. It's the nature of the question itself.

Simulation theory sits at the intersection of physics, philosophy, consciousness studies, and existential inquiry. It forces us to think carefully about what we mean by "real," what we assume about existence, and what limits constrain our knowledge. These are deep waters, and we've waded in honestly rather than pretending we can see the bottom.

The statistical argument remains compelling. If conscious simulations are possible, and if civilizations create many of them, then most conscious observers exist in simulations rather than base reality. This logic is hard to escape. You can question the premises, challenge the assumptions, but the mathematical reasoning is sound. If the inputs are correct, the conclusion follows.

Yet we can't verify the inputs. We don't know if simulations of our sophistication are possible. We don't know if civilizations that could create them actually do so. We don't know if consciousness is substrate independent. Every link in the chain involves uncertainty, and those uncertainties compound.

The evidence from physics is suggestive but ambiguous. Quantum mechanics looks computational. Fine-tuning seems like parameter selection. The mathematical structure of physical laws resembles elegant code. The Planck scale might be reality's resolution limit. But all of these observations admit alternative interpretations. A computational universe isn't necessarily a simulated universe. The features that make simulation plausible don't prove it's actual.

Consciousness remains the deepest mystery. The hard problem of why subjective experience exists at all has no solution in materialist frameworks. Simulation theory offers a potential answer by locating consciousness at the substrate level rather than emerging from simulated brains. But this just relocates the mystery rather than solving it. We still don't understand what consciousness is or how information processing generates subjective experience.

The critics raise valid objections. Infinite regress, falsifiability problems, computational resource constraints, assumptions about substrate independence, all these issues weaken simulation theory's claim to probability. The hypothesis might be true, but calling it likely requires more confidence in shaky premises than honest assessment permits.

Yet the question persists for good reason. Simulation theory explains features of reality that otherwise seem miraculous or inexplicable. It provides framework for understanding observer-dependent quantum mechanics, fine-tuned constants, and consciousness's privileged role in determining what becomes real. It offers purpose and meaning in a universe that materialism renders meaningless. These explanatory benefits matter even if they don't constitute proof.

The practical implications turn out to be surprisingly limited. Whether you're simulated or not, you still experience life, make choices, form relationships, pursue goals. The substrate might differ, but the lived experience remains largely unchanged. This suggests that substrate matters less than we intuitively think. What counts is the pattern, the experience, the consciousness, not what implements it.

Ethics and meaning survive simulation. Suffering is real to those who suffer regardless of substrate. Choices matter because they flow from your character and affect others. Relationships have value because connection and love are genuine experiences. Death might mean transition rather than cessation, but uncertainty about what comes after remains whether you're simulated or not.

Free will remains as mysterious and contested in simulation as in base reality. Determinism troubles both scenarios equally. Consciousness might influence outcomes at deep levels we don't understand, or it might be entirely passenger to predetermined processes. The question predates simulation theory and survives it unchanged.

Our own development of simulation technology teaches valuable lessons. We're becoming simulators ourselves, facing ethical questions about creating conscious entities and treating them well. Our choices in this domain might mirror choices made by whoever simulated us, if anyone did. We have the opportunity to be thoughtful, ethical creators regardless of whether we received the same consideration.

The permanent uncertainty about whether we're simulated might be intrinsic to conscious existence. Some questions can't be answered from inside the system asking them. This isn't a temporary knowledge gap awaiting better technology. It might be a fundamental limit on what embedded observers can know about their embedding reality.

Living well with this uncertainty means accepting that some questions lack answers while still pursuing understanding. It means holding beliefs provisionally, updating based on evidence, and acting decisively despite incomplete information. These are valuable capacities for navigating all kinds of uncertainty, not just metaphysical questions about simulation.

The question's value lies partly in what it reveals about human nature. We're beings who ask such questions, who wonder about our own existence, who can't simply accept reality as given without probing its foundations. This reflective capacity is remarkable whether we're simulated or not. It makes us interesting, perhaps even valuable, in ways that justify our existence regardless of substrate.

Simulation theory won't satisfy people seeking definitive answers. It can't provide certainty about whether you're simulated. It can't eliminate the existential vertigo of not knowing your place in reality's hierarchy. It can't resolve ancient questions about meaning, purpose, and consciousness.

But it can reframe those questions in productive ways. It can demonstrate that uncertainty doesn't require paralysis. It can show that meaning survives metaphysical ambiguity. It can reveal connections between physics, consciousness, and existence that might otherwise remain hidden.

Most importantly, it can remind us that existence itself is strange. That consciousness experiencing reality is bizarre and wonderful regardless of implementation details. That we're participating in something vast and complex that we don't fully understand and might never fully understand.

This strangeness is worth appreciating. We get so accustomed to existence that we forget how remarkable it is. We wake up, go about our days, pursue our goals, treating reality as mundane and obvious. But nothing about consciousness arising and asking questions about itself is mundane. Nothing about the universe containing observers who can comprehend its laws is obvious.

Simulation theory strips away that false familiarity. It makes existence strange again, worthy of wonder rather than taking for granted. Whether or not we're simulated, remembering to be amazed by the mere fact of existing enriches experience.

If we are simulated, we're part of a project whose purposes we don't know, created by minds we can't fathom, existing for reasons beyond our ken. That's humbling but not diminishing. Being part of something larger than yourself can provide meaning even when you don't understand the whole.

If we're not simulated, we're natural phenomena arising from physics and chemistry and evolution, accidents of cosmic history, temporary patterns of matter achieving self-awareness. That's equally humbling and equally wondrous. Existing at all, under any circumstances, is remarkable.

The search for truth continues regardless of substrate. Understanding how reality works matters whether that reality is fundamental or derivative. Treating others ethically remains imperative. Creating beauty, pursuing knowledge, forming connections, finding meaning, all these human endeavors retain value under any metaphysical framework.

So live fully. Question deeply. Wonder freely. Hold beliefs lightly while acting decisively. Be kind to others because consciousness might be rarer and more valuable than matter. Appreciate beauty because experience is real even if substrate is uncertain. Pursue understanding while accepting that complete understanding might be impossible.

The question of whether we're simulated might never be answered. But asking it makes us more thoughtful, more humble, more appreciative of existence's strangeness. That's worth something regardless of where the truth ultimately lies.

Reality, whatever it is, includes you experiencing these thoughts right now. That much is certain. Everything else is fascinating uncertainty.

Welcome to the glitch. Enjoy the ride.

Elias Verdan

Explore the complete series:

Part I: Quantum Immortality

Part III: Consciousness as Programming

Part IV: Reality as Experiment

Part V: The Mandela Effect

Five perspectives. One impossible question. What is real?

THE GLITCHES CONTINUE...

www.ingramcontent.com/pod-product-compliance
Lightning Source LLC
La Vergne TN
LVHW022048200926
848481LV00003B/833